Fire, Static and Dynamic Tests of Building Structures

Fire, Static and Dynamic Tests of Building Structures

Proceedings of the Second
Cardington Conference
Cardington, England
12–14 March 1996

Edited by

G.S.T. Armer

and

T. O'Dell

Routledge
Taylor & Francis Group

LONDON AND NEW YORK

First published 1997 by E & FN Spon

2 Park Square, Milton Park, Abingdon, Oxfordshire OX14 4RN
52 Vanderbilt Avenue, New York, NY 10017

Routledge is an imprint of the Taylor & Francis Group, an informa business

First issued in paperback 2019

ISBN 978-0-419-21680-3 (hbk)
ISBN 978-0-367-86370-8 (pbk)

A catalogue record for this book is available from the British Library

Publisher's Note This book has been prepared from camera ready copy provided by the
individual contributors.

CONTENTS

Views expressed in this document are those of the individual contributors and not necessarily those of BRE.

PREFACE

The Building Research Establishment (BRE) is the UK's leading source of authoritative research based advice on the design, construction and performance of buildings and on fire safety. It is a government agency within the Department of the Environment, tasked with assisting in the implementation of government responsibilities and initiatives related to construction.

The BRE's Large Building Test Facility (LBTF) is the most important single physical resource worldwide for the advancement of the understanding of whole-building performance. Most aspects of a building's lifecycle, from fabrication to fire resistance and explosions through to demolition, can be investigated on complete buildings. Full-scale buildings up to ten storeys high may be constructed in a protected environment allowing complete control over loading, test programming and data security.

This unique facility is already being used by industrial organisations, both as individual companies and sectoral groups, universities and research institutes, and government departments and agencies.

This book comprises the proceedings of the Second Cardington Conference. The principle topics covered relate to the programme of major fire tests undertaken on the eight-storey steel-framed composite bulding, forced vibration dynamic tests, and static load tests. New research programmes and buildings for the facility are in the planning stages and several papers are devoted to a description of the activities and the background to their development.

These proceedings are the only medium in which the full programme of work at the Large Building Test Facility is presented. The conference was the second in a series planned to provide a forum for users of the facility to publish their results and to exchange views on the way forward.

LBTF Steering Group
January 1996

PART ONE

BRE Fire Programme

FULL-SCALE FIRE TESTS ON COMPLETE BUILDINGS

D.B. MOORE
Structural Design Division, Building Research Establishment, UK

1 Introduction

A large proportion of research into the behaviour of structures has been concerned with the structural performance of isolated members, subassemblages, and the development of analytical techniques. Local failures in structures are generally studied by large-scale component testing, while the overall behaviour of the structural system is investigated using either scaled-down experimental models or subassemblages. Such tests are designed to provide basic data for model development and verification. However, these simplified tests cannot truly represent the behaviour of a complicated building fabricated and erected under normal commercial conditions, carrying floors, walls, cladding and many so called non-structural components.

So many differences arise in the behaviour of isolated components when they are connected together that questions concerning the force redistribution capability of highly redundant structural systems cannot be answered by component testing alone. Furthermore, the global and local failure behaviour of the building and the effectiveness of both structural and non-structural repairs can be proven only with tests on a number of different types of completed buildings.

In response to this need the Building Research Establishment developed, at its Cardington Laboratory, a facility for the full-scale testing of buildings up to ten storeys in height.

2 Experimental building

The first experimental building to be constructed within the Large Building Test Facility is a steel-framed structure representative of a modern office block (Figure 1).

Figure 1.   Steel-framed building

It is eight storeys in height and has a five bay long by three bay wide rectangular area of approximately 945 m². The building is designed as a no-sway structure with a central lift-shaft and two staircases providing the necessary resistance to lateral wind loads. The main steel frame is designed for gravity loads and the connections are designed to transmit vertical shear only. The floor construction is a steel deck and light-weight in-situ concrete composite floor. All structural components have been designed to the most up-to-date European Standards.

3 Fire tests

There is a growing opinion that the structural contribution of composite steel deck flooring systems is under-utilised in current design procedures, particularly for the fire limit state. This, together with evidence from three-dimensional numerical models and investigations from real fires such as occurred at Broadgate, suggests that the fire resistance of complete buildings is significantly better than that of the single elements from which fire performance is generally assessed. Furthermore, available mathematical models for the analysis of fire loaded buildings need to be verified against test data from real buildings before they can be used with any confidence.

Whilst it is possible to study structural behaviour by examining fire damaged buildings, interpretation of the findings is complicated by the lack of information on heating rates, temperatures, and stresses imposed on the members at the time of the fire. It is therefore desirable to undertake full-scale, compartment fire tests in real structures to quantify the structural contribution of composite steel deck flooring and the effects of non-structural components, and to quantify safety margins.

Until recently the scope for carrying out such tests was very limited, but the development of BRE's Large Building Test Facility (LBTF) at its Cardington Laboratory has given the construction industry a unique opportunity to carry out fire tests on a building designed and built to current practice.

A European collaborative programme of fire tests is being undertaken on the eight-storey steel-framed building. This programme is jointly co-ordinated by the Building Research Establishment and British Steel Technical and is supported by the European Coal and Steel Community and the Department of the Environment. Other organisations involved in the programme include TNO Building and Construction Research, Centre Technique Industrial de la Construction Metallique, The Steel Construction Institute and the University of Sheffield.

The complete testing programme is summarised in Table 1 and includes a range of fire tests from those on individual elements within the building, to fire tests on compartments representing large open-plan office accommodation. The Building Research Establishment has overall responsibility for tests 1, 2, 6 and 8. This report is concerned with tests 6 and 8 – representing a large office or conference room and a much larger open-plan office. In particular, this report presents the philosophy for carrying out such tests, gives the objective for each test and describes the construction of each compartment.

4 Philosophy

The main aim of these compartment fire tests is to assess the behaviour of the building under actions that correspond, as closely as possible, to the actual actions on a typical office building. This imposes two requirements on the tests: Firstly the test compartments should be identical to a real compartment in every essential detail, and secondly, the loading (both static and fire) should be representative of the actual loads found in a modern office block.

This approach is different to the more common and better understood laboratory tests. Traditional laboratory tests have attempted to isolate a single, or, at most, a small number of parameters for consideration by introducing idealised conditions. With the proposed fire tests the intention is, as far as possible, not to introduce such idealised conditions but to consider realistic scenarios both in terms of loading and of compartment design. The loading and compartmentation are therefore chosen to simulate scenarios which are likely to occur in the world of the built environment.

Further differences can be identified between the type of information generated by traditional laboratory tests and that generated from the Cardington tests.

Laboratory tests are used to calibrate analytical models, to validate predictive models and to evaluate bonding conditions. In such tests the conditions are tailored to represent the assumption sustaining a theoretical method, and the experimental data will usually be a simple parameter corresponding identically with the output data from the theoretical model. However, a single compartment fire test on the

Cardington building can generate different classes of information. For example, the large open-plan office fire test could generate data on the growth of fires in large spaces which could then be used to calibrate existing fire growth models; there are also a number of thermal models that predict temperature gradients through structural elements and the data from this test could be used to check these models.

Little is known about the response of large composite flooring systems to heating of this type and so a contribution to our knowledge of the structural behaviour of composite floors can be expected.   Lastly it will be a demonstration of the performance of a commercially-designed building in a realistic fire, so unlike traditional laboratory tests, one 'real' test can generate information of several different types so long as the observers have identified their needs and put in place the appropriate devices to generate the data. What cannot be assumed is that the data acquired for one purpose will be of any use for another.

The overall objective of the BRE fire tests is to observe and monitor the behaviour of both the structural and non-structural elements within a real compartment subject to a real fire. Construction of a 'real' compartment has proved to be relatively straight forward, but there has been much debate on what constitutes a 'real' fire. Much of the discussion has centred on whether or not glazing should be included in the compartment. On this topic there seems to be two schools of thought: Those who believe that the fracture and falling out of the glass is unpredictable because it is dependent upon such variables as the outside air temperature and the manufacture and fixing of the glass, and those who believe that the glazing has such an important influence on the development of the fire that to conduct a 'real' fire test without it (or to replace it with something more predictable) would seriously affect the credibility of the results.

The supporters of these two schools of thought could be seen as identifying two alternative, but equally valid, scenarios. A compartment with no glazing, or with a significant amount of the glazing removed, could be likened to a fire in a building under construction or a fire in a building on a hot summers day where the occupants have opened the windows. A fully glazed compartment is similar to a mechanically ventilated building. Reconciling these two different scenarios in a single test is almost impossible.

Two tests are therefore proposed for the BRE corner test (Test 6, Table 1), while a single test is proposed for the Large Compartment fire test (Test 8).

The first test on the corner compartment would be with all glazing in position and the second test would be with a significant amount of glazing removed. The main aim of the first test would be to investigate the behaviour of the glazing system and its influence on the development of the fire, while the aim of the second test would be to observe the development of the fire with a fixed initial ventilation condition and to examine its influence on the behaviour of the structure. Of course whether the second test goes ahead will depend on the outcome of the first test. In this regard the first test would seem to have the following possible outcomes:

1. The glass cracks, but does not fall out, and the fire dies. Very limited or no structural damage.
2. The glass cracks but insufficient glass falls out for the fire to fully develop. Very limited or no structural damage.

3. The glass cracks, falls out, and the fire develops to flash over. Significant structural damage.

Only outcome 3 would yield information on the behaviour of the structure.

Table 1. Cardington Frame Fire Test Programme

Test No.	Description	Date
1	Column tests, bare steel frame	Complete
2	Column tests, composite floors	Complete
3	Restrained beam	Complete
4	2D Plane frame	Complete
5	BST corner test	Complete
6	BRE corner test	Complete
7	Demonstration, furniture fire	May 1996
8	Demonstration, crib fire	April 1996
9	Column tests, failure	October 1996

The Large Compartment fire test will be glazed but a significant amount of the glazing will be removed to simulate the ventilation condition in a manually-ventilated building on a hot summers day. The main aim of this test will be to investigate the development of a fire in a large open-plan compartment, and to observe the effect of the fire on a large area of composite floor and the surrounding structure.

5 Static load, fire load and ventilation

A survey undertaken by the Building Research Establishment on serviceability loads in office accommodation established the actual loads to be approximately one third of the imposed load currently recommended in UK Standards. This is equal to a load of 0.83 kN/m^2 (2.5 kN/m^2 divided by 3). This, together with an allowance for the full weight of raised floors (0.4 kN/m^2), services (0.25 kN/m^2), false ceilings (0.15 kN/m^2) and partitions (1.0 kN/m^2) has been applied to the building using 1.1 kN sandbags. These sandbags have been distributed evenly on each floor to simulate as closely as possible a uniformly distributed load. The loads applied to the roof are higher and have a different distribution to those applied to the floors, due to the increased roof loads that would be caused by mechanical plant.

It is difficult to define accurately the fire load in a typical office because available data from surveys of fire load show enormous scatter. These variations are attributed to such factors as inherent national and/or regional differences, differences in the techniques of sampling and evaluation and differences in estimating that part of the fire load which is not consumed by the fire. However, it is generally accepted that 40 kg/m^2 is a reasonable design value of fire load for a typical office.

Estimating ventilation is also difficult because it depends on the area of broken glass, the area of open or burned-down doors or walls, and any additional ventilation routes a room may have with the rest of the building. For all the fire tests it is proposed to carry out ventilation tests to determine the base ventilation condition. A video recording will be taken of each test from which the area of broken glass can be estimated and the ventilation area calculated.

6 Corner tests

The size and location of these tests was a compromise between the following requirements:

1. The size and location of a real room in an office block.
2. The provision of a sufficient number of columns, beams and the floor they support to assess the performance of both structural and non-structural components and in particular the load shedding between the elements of the steel frame and the composite floor.

Fulfilment of the first requirement is considered to be essential to give credibility to the results, therefore an interior designer was commissioned to design floor layouts for all parts of the office building. A typical floor layout is shown in figure 2 and is based on occupancy by an international law from employing 318 people. It is appreciated that this is only one of a number of clients that could occupy the building, and that each client's requirements would be different, but it at least gives a starting point for deciding on the size and location of a typical room.

To investigate load shedding between the beams and the composite slab requires the slab to span at least one secondary member. A large area would clearly be a more severe test of this potential bridging mechanism. Nevertheless, it was considered that this small area would enable the validity or otherwise of this mechanism to be established. It would also act as a preliminary test to the much larger open-plan office test.

Based on the above requirements, the size and location of this test were chosen to represent a large office or conference room. The location for this test was the third floor of the building between grid lines E to F and 3 to 4, a room 9 m long and 6m wide. The location is shown in Figure 3 while Figure 4 is a three-dimensional view of the compartment.

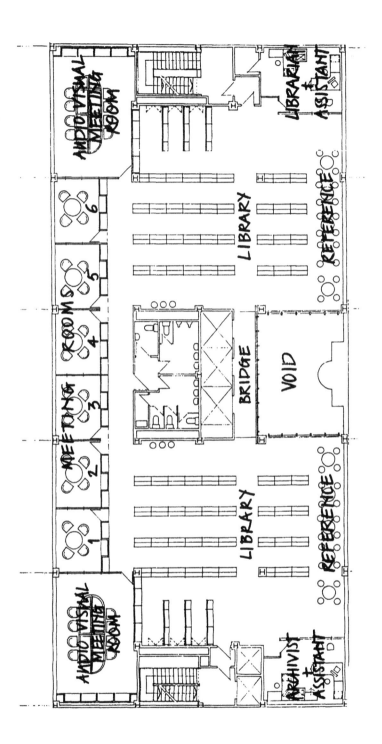

Figure 2.　A typical floor layout

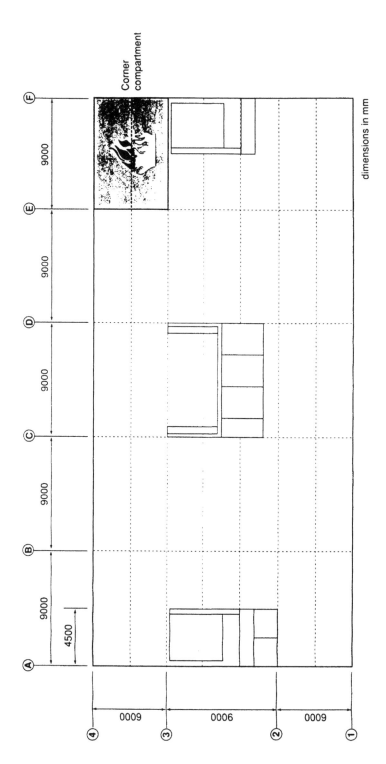

Figure 3. Corner fire test

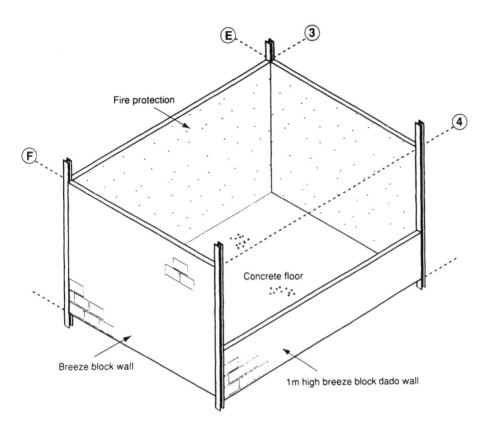

Figure 4.   3-D view of corner test

The proximity of the stairwell to the compartment required the walls of the compartment to be full height and fire resisting. All the steel beams within the compartment were left unprotected to investigate load shedding and bridging mechanisms between the beams and composite slab.

7 Large compartment test

The size and location of this test was again a compromise between the size and location of a typical open-plan office and the provision of a sufficient number of structural elements to assess load shedding between the main frame elements and the composite slab over a large area. Figure 5 shows the size of a typical open-plan office. The location of this test is the third floor of the building between grid lines A to C and 1 to 4.

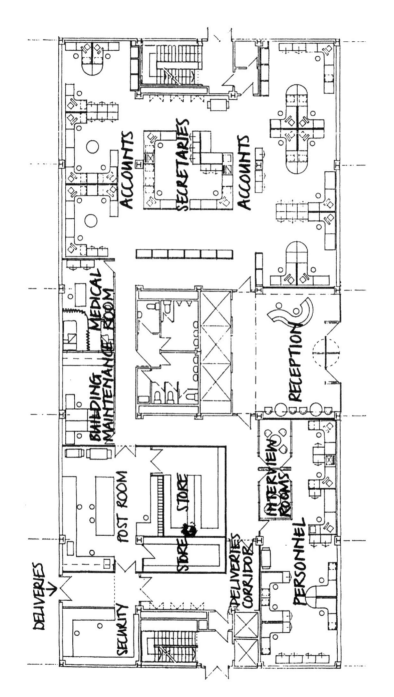

Figure 5. Typical open-plan office (Accounts department)

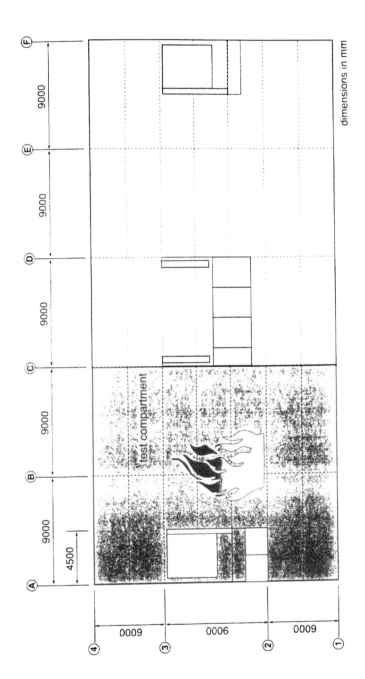

dimensions in mm

Figure 6. Proposed large compartment test

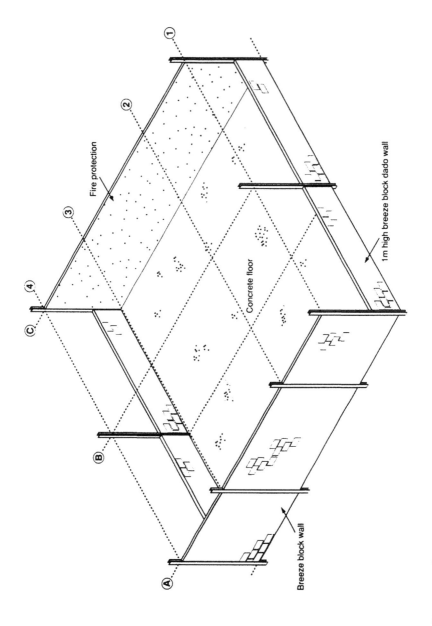

Figure 7. 3-D view of large compartment test

Figure 6 shows a plan of the room and Figure 7 shows a three-dimensional representation of the compartment. The room is 18 m long and 21 m wide and, again because of its proximity to the end stairwell and the central lift-shaft, it is bounded by full height fire resisting walls.

The main aim of this test is to investigate the ability of a large area of composite slab to support the applied loads once the main beams have failed. Thus all the beams, both primary and secondary, have no fire protection. All columns are fire protected to their full height.

8 Conclusion

This report presents the philosophy and main construction details of two full-scale fire tests on the eight-storey steel-framed building at BRE's Cardington Laboratory. These tests include a corner test representative of a large office or conference room and a much larger open-plan office.

The principle aim of these compartment fire tests is to assess the actual behaviour of a building and its components under actions that correspond as closely as possible to the actual actions on a typical office building. These tests will provide high quality data to develop the next generation of predictive models from which will be developed new, improved and safer recommendations for predicting the fire behaviour of real buildings.

A REVIEW OF COMPARTMENT FIRE TESTS TO EXPLORE THE BEHAVIOUR OF STRUCTURAL STEEL

G.M.E. COOKE
International Fire Consultant

Abstract
The fundamental complexity of modelling the severity of fully developed fires in a building has meant that reliance has always been placed in empirical relationships validated by fire tests conducted in compartments. This paper provides a historical perspective of small-scale and full-scale compartment fire tests conducted mainly in the UK over the past 30 years to determine the severity of fires and the performance of steel members exposed to them.

1 Introduction

A large portion of the building regulations in the United Kingdom have been, and still are, concerned with structural fire precautions. The need for economy in construction provides a powerful incentive for ensuring that fire regulations are not too onerous, but reduced levels of fire precautions arising from refinements in design will only be acceptable if it can be shown that any change in the levels of fire safety is socially acceptable.

Improvements in regulations and guidance documents can be best achieved by a rational scientific approach. However, before it is possible to determine the response of a structural member exposed to fire, it is necessary to be able to define the severity of the fire. Fortunately a number of early studies were made, mostly using model compartments for reasons of economy, to examine the factors affecting the severity of compartment fires. These studies [1, 2] concerned the physics of fire and explored, amongst other things, the relationship between ventilation and fire load on burning rate, and showed, for instance, that rate of burning for a ventilation-controlled fire was proportional to $A\sqrt{H}$ where A is the area and H the height of ventilation opening. However, model tests do not enable the response of structural members to be determined in a convincing way; tests needed to be carried out in full-scale compartments which could accommodate structural members representative of the sizes used in buildings.

In a building the ventilation opening is represented by windows when the glazing falls out, by open doors or other areas of low fire resistance.   The behaviour of glazing in fire is notoriously variable, and, to reduce the number of variables in experimental programmes of research into fire severity, it has been common practice to use unglazed openings, as is the case in almost all of the experiments described in this paper.   Such research programmes have also been confined to single compartments, whereas in practice a fire in one room or space may spread to another reaching their peak severities at different times.   When applying the results of this research these assumptions might be important and need to be considered.

Over the past three decades the steel industry in the UK has been particularly active to show, by the application of fire science and engineering, that structural steel-framed buildings are able to withstand the effects of fire without putting lives at risk or leading to high property losses.   The cost effectiveness of steel-framed buildings has steadily improved and the market share has increased, especially in multi-storey buildings.   The industry's strategy has been to obtain comprehensive information on:

1. the fire resistance ratings of a range of individual members such as beams, columns and floors exposed to the standard fire (as defined in BS476 Parts 8 and 20, and ISO 834)
2. heating rates for protected and unprotected steel sections, and, more recently, the load carrying performance of sub-assemblies and complete frameworks when exposed to a range of natural fires of different severity in compartments.

Since the mid-1960's carefully planned full-scale compartment fire tests have played an important role in the steel industry's strategy to increase the usage of steel in buildings. The paper now reviews a number of milestone experimental programmes which, in chronological order, had the following aims:

- to quantify fire environment
- to quantify fire environment and thermal performance of non-loaded individual steel members simultaneously
- to quantify fire environment, thermal and structural performance of a loaded steel sub-assembly simultaneously
- to quantify fire environment, thermal and structural response of complete loaded steel structures

Most of the experiments involving steel were jointly undertaken by Fire Research Station and the public sector steel industry (initially the British Iron and Steel Federation, then British Steel Corporation and finally British Steel). Funding came from the Department of the Environment, the steel industry and, most recently, the European Coal and Steel Community (ECSC).

2 CIB model compartment fire tests

In 1958 the then Director of the Fire Research Station recognised there was a paucity of available information on fully-developed fires, and that the problem was world-wide. He proposed an international programme of fully-developed fire tests in model scale single compartments. The experimental study was undertaken under the auspices of the Conseil International du Bâtiment (CIB). More than 400 statistically designed experiments shared by eight laboratories were made using model compartments constructed of asbestos millboard with external steel framing. The fire load comprised standardised timber cribs built up from square section sticks. Cribs were used because they provided a reproducible fuel bed whose properties could be varied systematically. A wide range of parameters were examined:

1. *Fire load densities.* Representative of the range of fire loads encountered in normal occupancies (excluding warehouses). They were mainly 20, 30 and 40 kg/m^2 but in a few tests 10 kg/m^2 were used
2. *Fuel thickness and spacing.* The wood sticks were 10, 20 or 40 mm thick and the horizontal spacing between sticks was 0.3, 1 or 3 stick thicknesses. The sticks were glued together. This allowed different burning rates to be considered.
3. *Shape and scale of compartment.* Four shapes of compartment were used in still air conditions: 211, 121, 221 and 441 where the first figure is width, the second depth and the third height. The scale of the compartment was taken as the compartment height: scales of ½, 1 and 1½ were used.
4. *Ventilation opening.* This was rectangular, extending from floor to ceiling and occupying ¼, ½ or full-width of the front of the compartment. In a few of the tests the opening was ¾ of the front.
5. *Wind speed and direction.* The Fire Research Station facilities were used for wind speeds of nominally 5 and 8 m/s for wind blowing at right angles to the plane of the ventilation opening and at 60° to this direction.

Measurements included:

- the loss in weight of fuel from which the rate of burning could be derived
- the temperature of thermocouples placed within the compartment;
- the intensity of radiation at a point in front of the ventilation opening, which could be used in a heat balance and give a measure of the radiation hazard to neighbouring buildings;
- the intensity of radiation from the flames above the ventilation opening at a point close to the plane of the front of the compartment, which is a measure of radiation hazard to the floor above the fire.

The main conclusions of this work [3] were as follows:

1. The conventional relationship for maximum rate of burning $R = kA_wH^{\frac{1}{2}}$ (kg/s) was a gross approximation, where k = a constant, usually taken as 0.09, A_w = area of opening, and H = height of opening. The CIB work suggested that k could approach double this value for large area compartments.

2. The intensity of radiation from the ventilation opening was related to the rate of burning per unit opening area and the temperature within the compartment.
3. The intensity of radiation from the flame could be related to the rate of burning, the intensity of radiation at the ventilation opening, and the dimensions of the compartment.
4. For the same fire load per unit area of opening, higher fire resistance was required in shallow (eg cubical) compartments than in deep or elongated compartments.
5. A simplified calculation of the fire resistance requirement for a structural member showed that the fire resistance t_r was given by:

$$t_r = b \times \text{total fire load} / [\text{ventilation area} \times (\text{wall} + \text{ceiling area})]^{\frac{1}{2}}$$

6. where b varied from 1.1 to 1.3 for the model experiments and was 0.95 for a wide range of large-scale experimental fires. This correlation largely eliminated the effect of compartment shape.

3 The UK steel industry compartment fire programmes

The steel industry's work began in late 1963 with a single fire test in a specially constructed brick building, representative of a Local Authority flat furnished with domestic furniture. A one-hour fire resistance was required, according to the building regulations, and this resulted in the use of costly fire protection. The test showed [4] that the required fire resistance was of the order of 20 minutes, to keep steel temperatures below the critical value of 550 °C.

However, the result of a single test of this kind was considered an unsatisfactory basis on which to approach the regulatory authorities for a change. Information was needed on the effect of a wider range of parameters before this was possible.

3.1 The BISF/JFRO compartment tests in the mid 1960's.

The steel industry (the then British Iron and Steel Federation (BISF)) and the Fire Research Station (the then Joint Fire Research Organisation (JFRO) of the Ministry of Technology and the Fire Offices' Committee) sponsored a comprehensive programme of fully developed compartment fire tests in a purpose made building at Borehamwood. From the steel industry viewpoint the programme had three aims:

1. to examine the behaviour of fire, i.e. the factors which determine its severity and from this to identify the circumstances or position in which structural steel members could be safely used
2. to quantify the behaviour of steel in the various fire conditions, again so as to identify the safer situations
3. to compare the standard fire resistance test with real fire situations to see if there was any scope for amending the statutory fire resistance requirements

In this work [5, 6], which was probably the largest and the most comprehensive undertaken at the time, twenty-six fire tests were carried out in a specially constructed building using several conditions of ventilation and a variety of fire load densities typical of buildings in the low fire load class (e.g. residential and office buildings). Parameters known to affect fire severity were chosen as shown in Table 1.

Table 1.   Parameters affecting fire severity

Parameter	Values used
Nature of fuel	Cribs: using softwood sticks 45 mm^2 section, 1.1 m lengths. Fibre insulating board linings to walls and ceiling. Petrol and kerosine. Domestic furniture: mixed, medium and heavy.
Amount of fuel (fire load densities expressed as wood equivalent)	Cribs: 7.5, 15, 30 and 60 kg/m^2 Fibre insulating board linings, petrol and kerosine: 7.5 kg/m^2 Furniture: 15 and 25.5 kg/m^2
Arrangement of fuel	Cribs: covering $^1/_3$ and $^2/_3$ floor area. Fibre insulating board: uniformly disposed on walls/ceilings. Petrol and kerosine: eight trays, each 0.9 m square. Furniture: typical of domestic occupancy.
Size and shape of room	7.7 m x 3.7 m x 3 m high single compartment, or 7.7 m square (double) compartment.
Window area and shape	Two openings, 3.05 m x 1.83 m high in each of two opposite long walls such that single and double sided ventilation obtained with single and double compartments respectively. Each window width varied to give 1.3, 2.8 and 5.6 m^2 of opening corresponding to $^1/_8$, $^1/_4$ and $^1/_2$ ventilation.
Thermal insulation of room	0.34 m cavity brick external walls and lightweight concrete internal wall rendered with vermiculite plaster. Concrete floor. Ceiling of refractory concrete. With and without mineral wool slab 25 mm thick lining to walls/ceiling.
Glazing	Used only in some of the furniture tests.

Unloaded structural steel members (29 steel beams and columns) were arranged in, and outside, the fire compartment to provide the data concerning a variety of situations (Table 2), and their behaviour (i.e. their temperature) was observed and compared with appropriate fire resistance test data.

In most of the this work the fuel used was wood sticks arranged as cribs, but several tests were performed in which other fuels were used, for instance ordinary domestic furniture [7, 8].

The time-temperature curves for the tests using furniture are compared in Figure 1 with those for the wood cribs using a similar fire load density.

Table 2.   Parameters affecting behaviour of steel

Parameter	Internal Steelwork	External Steelwork
Weight/shape of steel section	37, 48, 52 and 107 kg/m – I section 18.75 and 43.8 kg/m – rectangular hollow section	37 and 52 kg/m – I section.
Type/number of elements	4 beams, 17 columns	3 beams, 5 columns
Location of element	Columns: built into wall, standing against wall and free-standing in compartment. Beams: built into wall, spanning compartment just below ceiling	Columns: against façade and 0.46 m away (unshielded column only) Beams: along top of window opening
Area of element heated	Maximum (free-standing), to Minimum (fully built in)	Max. (centre window opening) Min. (shielded by brickwork)
Loading & restraint	None applied – concerned only with thermal behaviour	
Protection (encasement)	Nil 13 mm mineral wool slab 19 mm asbestos insulating board 25 mm tongued and grooved softwood	Nil

These indicate that the particular wood cribs chosen as a convenient and easily reproducible fire load for test purposes, are also representative of furniture. The important comparison is the maximum temperatures reached, not the time taken to attain those temperatures. Some conclusions were as follows:

- The higher the fire load density, the higher the temperature reached in the fire. The indications are that this effect will level off at the higher fire load densities, and that the maximum average spatial temperature reached will be 1200 °C.
- At each fire load density, the lower degree of ventilation (i.e. the smaller window opening) resulted in both a hotter fire and one of longer duration.
- The consequential conclusion that fire load density is not a complete guide to the severity of a possible fire is in agreement with information already obtained from the CIB model experiments at the Fire Research Station.
- For the two lowest fire load densities, 7.5 kg/m^2 and 15 kg/m^2 with both degrees of ventilation, practically no flames emerged from the window openings, but some knowledge of the behaviour in other compartment shapes and of other fuel dispositions was needed before this can be regarded as a completely general statement.

- Where the external steel members were in such a position that flames could not impinge upon them, they did not reach a temperature higher than 300 °C, even for the most severe condition used (i.e. for a fire load density of 60 kg/m^2 with the lower ventilation). However, where flames could reach the steel members, the critical temperatures of 550 °C was quickly reached. The advantage of placing steelwork away from the facade was clear.

- Large differences in temperature were recorded in the external steel members (as between the front and back flanges, for instance) and this gave rise to temporary distortion of the members owing to differential thermal expansion.

- For fire tests of the lowest fire load density used, i.e. 7.5 kg/m^2 with both degrees of ventilation, the temperatures reached by the internal unprotected structural steel were well below the critical temperature, the maximum temperature recorded in these instances being 370 °C.

- With the fire load density of 15 kg/m^2 and ½ ventilation, a temperature of 510 °C was recorded on one point on the flange of the free standing internal steel column, but the maximum of the average temperatures of the column was 387 °C. No test was carried out in which internal unprotected steel columns were exposed in a fire of fire load density 15 kg/m^2 with ¼ ventilation, but it would be reasonable to expect that the temperatures reached would equal, if not exceed, the critical value of 550 °C.

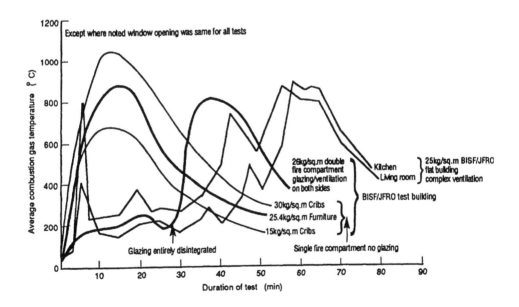

Figure 1. Average combustion gas temperatures in the BISF/JFRO test buildings showing correlation between crib/furniture fires and the effect of glazing.

Hence the use of unprotected steel inside a building appeared difficult to justify, but the work showed that in fires with a low fire load density, particularly with large areas of window opening, structural steel would remain below its critical temperature. The size of such fire load density is, however, lower than those met in offices, housing and hotels.

The results of this work have been comprehensively analysed in terms of the parameters determining the severity of fire [9], the heat transfer to the steel [10] and a comparison between furnace tests and the compartment fires [11]. A brief review of the work, aimed specifically at the steel industry, is also available [12].

3.1.1 Tests on unloaded steel members

A collaborative test programme was initiated jointly by the Swinden Laboratories of the British Steel Corporation (BSC) and the Fire Research Station of the DOE Building Research Establishment to determine the heating rates of a wide range of unloaded and unprotected steel beams and columns in fires of different, but known, severity. It may be recalled that the earlier programme of tests by BISF/JFRO had involved 11 unprotected members and 17 protected steel members, with the unprotected members all being roughly the same size. The new programme sought to examine the effects of fire on a larger range of unprotected section size, and highlight situations where unprotected steelwork had adequate inherent fire resistance. Low fire loads and large ventilation opening were selected to represent occupancies such as schools and multi-storey office blocks. A compartment measuring 8.6 m x 5.5 m x 3.9 m high was built at the Cardington LBTF, and a total of 21 fire tests were performed which were independent of weather conditions. Thermocouple data on the time dependence of combustion gas temperature and steel temperature were collected.

Eight tests compared temperature-time data for timber crib fire load densities of 10, 15 and 20 kg/m^2 using $^1/_2$, $^1/_4$, $^1/_8$ ventilation of one wall (the long wall) of the compartment. Four timber crib fires examined the effect of altering the ventilation for a given opening factor, the effect of an unevenly distributed fire load, different methods of ignition and a change in the compartment lining. A further five tests examined the effect of incorporating polystyrene with the timber fire load. The main conclusions of the work [13] were:

- an increase in fire load density and a reduction in ventilation both raised the maximum combustion gas temperature attained. The combustion gas temperature-time curve in the early stage for a fire of 15 kg/m^2 and ¼ ventilation was similar to that of the BS476: Part 8 (ISO 834) standard furnace curve.
- the wood/polypropylene fire resulted in a rapid rise in combustion gas temperature in the early stage of a test, and was accompanied by large volumes of block smoke, but had little effect on the maximum steel temperatures attained.
- the temperature rise of the steelwork was clearly affected by the section factor (the fire-exposed perimeter divided by the cross section area) and its position within the compartment.
- For a given fire load density and ventilation opening, the peak combustion gas temperature was highest for a uniformly distributed fire load simultaneously

ignited, lower for an unevenly distributed fire load simultaneously ignited, and lower again for a uniformly distributed fire load ignited at one point and allowed to spread naturally.

- the water cooling of a rectangular hollow steel section was effective once recirculation problems had been solved. The steel temperature was then kept below 220 °C.
- temperature gradients across the section of columns partially built into the double-leaf external wall reached 400 °C and resulted in substantial thermal bowing. Computed deflections agreed well with measured valves [14].
- The effect of changing the compartment lining, from insulating refractory brick walls and ceramic fibre ceiling tiles to Gyproc 'Fireline' plasterboard walls with the removal of the ceiling tiles, markedly lowered the combustion gas temperatures. This confirmed the need to take account of the thermal properties of the enclosure in any analytical approach (e.g. the time equivalent method) for the calculation of fire resistance.
- the investigation provided valuable experimental data on the behaviour of nature fires and the behaviour of unprotected steel. The data enabled analytical methods to be used for the design of steel structure avoiding the need for expensive and time consuming fire tests.

3.1.2 Test on a loaded steel frame.

All of the tests reported so far were on individual unloaded members. It was believed that the performance of a frame in fire would be better than its individual members and it was therefore decided to conduct a fire test on a full-size, fully loaded, two-dimensional, mainly unprotected steel frame, in the 8.6 x 5.5 x 3.9 m high compartment described above.

The steel framework, two columns nominally 3.5 m and a beam nominally 4.5 m long, was typical of a frame used in a 2 or 3 storey building. The beam was unprotected but the columns had their webs protected by autoclaved aerated concrete blocks. The test arrangement is shown schematically in Figure 2. Identical members had previously been exposed in the standard furnace so that their individual fire resistances were known. The compartment fire test showed [15] that the performance of the frame was better than that of the individual elements, which was mainly attributed to beneficial beam/column action on the beam, and the effect of blocking in the web of the column.

A numerical simulation of this test has been made [16] which examines the influence of lateral restraint, frame continuity and thermal expansion.

3.2 The BSC/FRS large compartment tests in the early 1990's

This experimental programme was undertaken urgently so that the DOE Construction Policy Directorate would be able to decide whether or not to accept an equation for equivalent time of exposure contained in the Structural Eurocodes. The equation permits the fire resistance, needed by a structure to survive a burn-out of the contents, to be calculated in terms of fire load, ventilation area and the thermal insulation properties of the enclosing walls, floor and ceiling. If the equation underestimated the real fire severity, this could lead to constructions that were unable to contain the fire – a potentially dangerous situation.   At present the levels of fire resistance stipulated under Building Regulation guidance documents are not related to ventilation or thermal properties, and include various factors of safety (for high-rise buildings for example).   So there was concern that the Eurocode approach might give very different results.

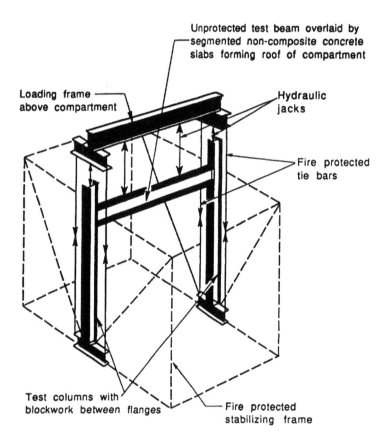

Figure 2.   Schematic layout of the loaded frame test rig used in a Cardington compartment test.

The time-equivalent equation had previously been validated by tests in small compartments (compartments having a floor area not greater than 60 m^2), but would it be valid for compartments having a larger floor area and larger depth-to-height ratio representative of, for instance, large open plan offices? To answer this question a large compartment (23 m deep, 6 m wide and 3 m high) was built with a ventilation opening at one end. It was designed to represent a 'slice' through a much larger compartment. The programme was undertaken by FRS and British Steel Technical and was sponsored by DOE and British Steel Sections, Plates and Commercial Steels. Nine experiments were made to [17] examine the effect of:

- varying the fire load density from 20 to 40 kg/m^2
- varying the size and shape of the ventilation opening in one wall, including a simulation of a poorly ventilated basement. Ventilation areas were 100, 50, 25 and 12.5% of the 6 x 3 m wall.
- changing the thermal properties of the walls and ceiling
- simultaneous ignition of all 33 cribs, compared with ignition of 3 cribs at the rear of the compartment, followed by natural fire spread to the others.

The concrete compartment was lined internally with a 50 mm layer of highly insulating ceramic fibre, and the concrete floor was insulated with 100 mm thick layer of dry sand. The fire load was in the form of 33 cribs, each 1m square arranged in 11 rows of 3. For reasons of safety the cribs were ignited from outside the compartment using fuse cord and incendiary devices. Six of the cribs in the central row were supported on load cell platforms so that weight loss could be measured.

Thermocoupled sections of unprotected and protected steel were suspended below the ceiling so that temperature data could be collected and compared with the temperatures attained by identical sections exposed in the standard fire resistance test. Other measurements included combustion gas temperatures, velocity of air flowing in and hot gases out of the compartment, radiation intensity, crib mass loss and temperatures of flames/hot gases emerging from the opening. The following conclusions were reached:

- Measured and calculated values of time-equivalents agreed well for most of the tests, provided that specific values of thermal insulation of the compartment linings were used in the calculation – the Eurocode gives insufficient data in this respect.
- The equation is inappropriate for long duration, poorly ventilated fires such as those experienced in basements, and underestimates the time-equivalents for some fully ventilated fires.
- The use of plasterboard as a lining material, instead of a ceramic fibre material of greater thermal insulation, led to a lowering of the combustion gas temperatures as might be expected.
- The duration of a fully developed fire can be large for a compartment having boundaries of high thermal insulation and restricted ventilation: a duration in excess of 8 hours was obtained with a fire load density of 20 kg/m^2 and ventilation of $^1/_8$ of the 6 m x 3 m front wall (Figure 3).
- Neither the temperature of the combustion gases (Figure 3), nor the heat transfer was uniform throughout the compartment.
- The mode of fire spread through the 23 m depth of compartment was interesting. When only the cribs at the rear were ignited the fire spread to the front via the

tops of the other cribs by spontaneous ignition, assisted by radiation from the layer of hot gases under the ceiling.   The front rows of cribs then burnt downwards and the next row similarly, while burning in the rear of the compartment stopped because all of the air was being consumed by the cribs at the front of the compartment.   Full depth burning of the remaining cribs then proceeded towards the rear of the compartment.   This mode of fire spread, from the front to rear, also occurred when all 33 cribs were simultaneously ignited.

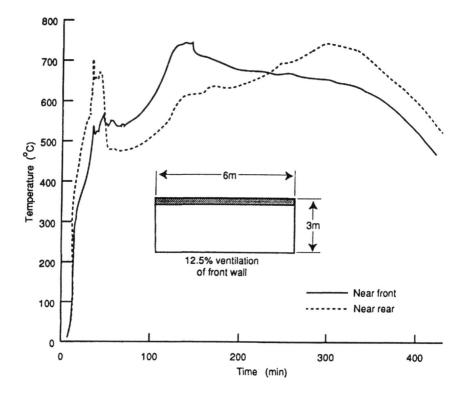

Figure 3.   Combustion gas temperatures for BSC/FRS large compartment test showing non-uniformity of temperature.

The effect of ventilation while keeping other parameters constant, Table 3, has again underlined its important effect on fire severity [18].

Table 3.   Effect of ventilation on required fire resistance

Ventilation area as a proportion of front wall area (%)	100	50	25	12.5
Measured equivalent time of fire exposure (mins)	71	80	99	111

For fire load density = 20 kg/m^2

The comprehensive data collected in the test programme will be of great use to fire modellers since there are few data for large compartment fires.

3.3 The BST/ECSC/BRE programme of tests on an eight-storey structure

For many years there has been anecdotal evidence which suggests that the fire behaviour of a framed structure was better than the behaviour of its individual members. This had been partly quantified by the loaded frame fire test described earlier [15] but this test was not designed to explore membrane action between beams and floor. However, the occurrence of a severe fire in the partially constructed 14-storey office building in Phase 8 of the Broadgate Development in the City of London, June 1990, and the following assessment of the fire [19] proved beyond doubt that composite action between composite steel deck floors and beams, and between beams and columns, was highly beneficial.

Shortly afterwards plans were made to construct an eight-storey steel-framed structure with composite steel deck floors at the BRE's Large Building Test Facility (LBTF) at Cardington to examine its response to a wide range of tests including static and dynamic response, as well as fire tests. This test structure would be 33.5 m x 45 m x 21 m high, and be used to:

- provide further information on natural fires in compartments within a real multi-storey steel-framed building, and assess the behaviour of the structural members including the floors
- validate computer models for the behaviour of the structure exposed to fire
- provide design guidance, using validated computer models, taking account of beneficial frame/floor interaction to quantify safety margins and provide a further check on the validity of the Eurocode equation for equivalent time of fire exposure

Two separately funded, but complementary, programmes of fire research were arranged and some of the early tests have been briefly reported [20].

3.3.1 Pan-European test programme

The experimental programme is funded by British Steel Plc, TNO, CTICM and the European Coal and Steel Community (ECSC), and is coordinated by British Steel Technical and BRE.

Tests were conducted on a restrained beam, a 2-dimensional plane frame and a 3-dimensional frame, and were conducted without glazing in the window openings. A further 'demonstration' test is to be made in which furniture is the fuel.

The restrained beam test was made on an unprotected secondary beam with shielded end connections, and a 3 m width of composite slab. The test, which ran for approximately 2¾ hours, indicated that the good result was mainly due to the restraining action of the surrounding structure.

The 2-D plane frame test was made on a 2.5 m wide strip across the full 21 m width of the structure using a gas fired furnace capable of supporting the floor should it fail. The protected column (two internal and two external) and three unprotected primary composite beams were enclosed by the furnace, and although the maximum exposed steel temperature was approximately 820 °C, the mid-span deflection of the primary beams did not exceed span/34. Further details are available [21]. Information on the 3-D frame test is not yet available.

3.3.2   BRE test programme

The BRE test programme involves: a series of tests on isolated columns heated using portable gas-fired furnaces [22] culminating in a test to failure; a demonstration test using cribs with glazed openings in a 21 m x 22 m compartment [23]; and a 3-D corner test.

Information on the BRE tests is not yet available as the programme was planned to commence once the pan-European tests were completed.

4   Conclusion

This review of compartment fire tests, conducted mainly in the UK, has shown that a considerable amount of experimental work has been undertaken to indicate the conditions under which unprotected structural steel members may be safely used. There are clear indications that the behaviour in fire of complete frameworks is better than that of individual members, and there are expectations that validated computer models will be able to show where refinements in design can be made.

5   References

1.   Thomas P.H., *Studies of fires in buildings using models: Part 1 Experiments in ignition and fires in rooms*, Research, Vol. 13, February 1960, pp 69-77.

2.   Thomas P.H., *Studies of fires in buildings using models: Part 11 Some theoretical and practical considerations*. Research, Vol. 13, March 1960, pp 87-93.

3.   Thomas P.H. and Heselden A.J.M., *Fully developed fires in single compartments*. A co-operative research programme of the Conseil International du Bâtiment (CIB Report No. 20), Fire Research Note No. 923, August 1972.

4.   Ashton, L.A., Report of special investigation on test to determine the effect of fire in a furnished flat on the protected structural steel frame. F.R.O.S.I. No. 3357, July 1964. Joint Fire Research Organisation.

5.   Butcher E.G., Chitty T.B. and Ashton L.A., *The temperatures attained by steel in building fires*, Fire Research Technical Paper 15, London, 1966, HMSO.

6.   Butcher E.G., Bedford G.K. and Fardell P.J., *Further experiments on temperatures reached by steel in building fires*, Paper 1 of Symposium No. 2 Behaviour of structural steel in fire, London, 1968, HMSO.

7.   Butcher E.G., Clark J.J. and Bedford G.K., A fire test in which furniture was the fuel, Joint Fire Research Organisation, Fire Research Note 695, 1968.

8.   Theobald C.R., and Heselden A.J.M., *Fully developed fires with furniture in a compartment*, Joint Fire Research Organisation, Fire Research Note 718, 1968.

9.   Heselden A.J.M., *Parameters determining the severity of fire*, Paper 2 of Symposium No 2, Behaviour of structural steel in fire, London, 1968, HMSO.

10.   Law Margaret, *Analysis of some results of experimental fires*, Ibid Paper 3.

11.   Butcher E.G. and Law Margaret, Comparison between furnace tests and experimental fires, Ibid Paper 4.

12.   Butcher E.G. and Cooke G.M.E., *Structural steel and fire*, Conference on Steel in Architecture held on 24-26 November 1969, Published by British Constructional Steelwork Association in 1970, London.

13.   Latham D.J., Kirby B.R. and Thomson G., *The temperatures attained by unprotected structural steelwork in experimental natural fires*, Fire Safety Journal, 12 (1987) pp 139-152.

14.   Cooke G.M.E., *The structural response of steel I-section members subjected to elevated temperature gradients across the section*, PhD thesis, The City University, London, September 1987, pp 114-117.

15.   Cooke G.M.E. and Latham D.J., *The inherent fire resistance of a loaded steel framework*, Steel Construction Today, 1987, 1, pp 49-58.

16.   Franssen J.M., Cooke G.M.E., and Latham D.J., *Numerical simulation of a full scale fire test on a loaded steel framework*, Journal of Constructional Steel Research, Vol. 35, No 3, 1995, pp 377-401.

17.   Kirby B.R. et al, *Natural fires in large compartments* - A British Steel Technical and Fire Research Station Collaborative project, British Steel Technical, Swinden Laboratories, June 1994.

18.   Cooke G.M.E., *The severity of fire in a large compartment with restricted ventilation*, IMAS '94 Conference: Fire safety on ships - developments into the 21st century, 26-27 May 1994, Institute of Marine Engineers, London.

19.   Steel Construction Industry Forum, *Investigation of Broadgate Phase 8 fire*, Publication No. 113, Steel Construction Institute, 1991.

20.   *First results from the Large Building Test Facility*, Proceedings of the 1st Cardington Conference 16/17 November 1994, Building Research Establishment.

21.   Kirby B. and Millsom, *ECSC fire tests update*, Cardington Large Building Test Facility Newsletter, Issue No 9, Building Research Establishment, 1995.

22.   Lennon T., *Fire engineering portable furnaces*, New Steel Construction, Vol. 3, No 1, February 1995, pp 21-24.

23.   Lennon T., *Large compartment fire test*, Proceedings of 1st International Conference on fire safety of large enclosed spaces, 25-27 September 1995, Lille, France, pub'd by Independent Technical Conferences, Kempston, Bucks.

HEAT FLUX MEASUREMENT IN REAL AND STANDARD FIRES

D.J. O'CONNOR, B. MORRIS and G.W.H. SILCOCK
Fire Engineering Research Centre, University of Ulster, UK

1 Introduction

The quantification of thermal impact, incident upon a structural element, is vital in order to establish the effect of fire severity on thermo-structural response. Observing that the direct effect of a fire environment arises as a heat flux incident upon to the surface of the structural component, the main objective of the research study described here is to assess the fire severity within the various large compartment fires proposed within the Cardington fire programme by means of the measurement of flux–time variation. In such a way, the performance of these 'natural' compartment fires can also be compared with the fire performance of standard fire test furnaces, mainly at the University of Ulster.

The measurement of heat flux is not a trivial issue and is currently determined using the Gardon gauge which is cumbersome to operate, expensive and regarded as not particularly robust. In the present study, another type of flux measurement device, the steel billet, is used. It is operated alongside the Gardon gauge for comparison purposes. In addition, it is intended that more of these devices will be allocated throughout the area of larger compartment fire tests in order to expand the field of data collected. This device is also being assessed independently in a laboratory programme utilising constant flux and standard fire test scenarios.

The flux-time measurements produced by the billet devices are also being compared with appropriate gas temperature measurements, obtained using both standard bead thermocouples and the plate thermometer as proposed by Wickström [1].

Expertise in flux measurement was initially gained during an EPSRC funded project [2] investigating the utility of a fire test methodology for half scale models, where the quantification of incident heat flux in standard fires was a fundamental issue. The current project on heat flux measurement [3] is sponsored by EPSRC with support from British Steel and the Building Research Establishment.

2 Background

Difficulties have been experienced in defining and measuring fire severity systematically, firstly in relation to the specification of a standard thermal flux insulation onto surfaces during furnace fire tests and furthermore, relating the flux created by a naturally occurring fire to that generated during a standard fire test. The use of local gas temperature measurement, as a method of interpreting the thermal environment causing thermal flux impact onto the structure under test, is recognised as being the only pragmatic solution for control of standard furnace fire tests or indeed monitoring fire development in wood crib fuelled compartment fire tests. Simplistically, the gas temperature measurement and control is crude in relation to the direct thermal flux onto the test specimen. In this regard, the complexity of surface thermal energy transfer processes is well recognised. In addition, particularly in standard test furnaces, there are other factors, which influence the temperature measured by control thermocouples and hence the actual thermal insulation produced by the overall furnace environment.

The current European harmonisation debate is testimony to this uncertainty. These problems are further exacerbated in relation to the measurement of fire severity in large compartments or naturally occurring fires, where the variability of fuel source, whether in the form of wood cribs or consisting of more natural sources like furniture, and compartment geometry make the dynamics of fire growth and combustion products less easily defined and consequently difficult, if not impossible to predict with any certainty.

The uncertainties described above also have implications for computer modelling of structural behaviour in fire. Presently, thermal analysis is effected by applying global surface heat transfer coefficients to fire gas and surface temperatures and the lack of agreement of rational values worldwide, for furnace emissivities in particular, highlights the problems in providing repeatable thermal analysis using consistent input parameters. If computer modelling is to achieve a useful role in fire engineering design an appropriate standard definition of fire severity must be agreed for standard and real fire scenarios.

The definition of fire severity in terms of an incident heat flux-time curve is an appealing proposal, which would be particularly useful in computer modelling. The application of this premise to primary control in fire testing hardly seems at present to be practicable. However, it is feasible to measure flux-time in fire tests in order to effect a comparative assessment of the actual thermal flux onslaught produced by the fire environment.

The heat transfer problem has received much attention in Europe over the past 5 years. In particular Wickström [1] proposed the plate thermometer as an alternative furnace control device and there has been a continued critique of this issue. The current European-wide round-robin program shows the interest in comparative assessment of furnace severity and the use of the 'SP Boras test panel' [4] indicates the need for calibration elements, which can measure the actual thermal onslaught on a structure in fire test situations. Cook [5] used a similar panel to carry out an assessment of three European furnaces and he also employed various devices [6] to assess fire severity in recent large compartment tests. Similarly, Kirby [7] continues to be successful in the use of thermocoupled steel sections as 'indicatives' to provide comparative assessment of compartment fires by correlating the data with measurements on similar specimens subject to standard fire test environments.

Other fires of 'calibration elements' have been proposed by many authors [8–11] in attempts to assess fire severity in furnaces. All are based on a knowledge of the thermal diffusion process through a typical structural element, usually a concrete brick or a thick steel plate. The incident heat flux is inferred from an evaluation of the flux necessary to promote a known thermal effect in the sample. This effect is normally the measurement of an instantaneous surface thermal gradient or an evaluation of overall transient heat flow response. The feature of all devices is that they require an inverse thermal calculation [12] , which forces an indirect means of flux measurement based on a complex back calculation approach.

The steel billet utilised in the present research programme [13] is in essence another calibration element. However, the associated billet calibration programme and the direct comparisons with Gardon gauge output within each test are intended to provide quantitative information in the form of flux-time curves rather than comparative fire severity data.

3 Surface heat transfer

The net heat flux q_n incident on and through the boundary of a surface is given by the equation:

$$q_n = q_f - q_s = h(T_f - T_s) + (\sigma \varepsilon_f T_f^4 - \sigma \varepsilon_s T_s^4) \tag{1}$$

where the two terms in brackets represent energy transfer between the hot gases q_f at temperature T_f and the surface q_s at temperature T_s by convection and radiation respectively. Generally, convective transfer is considered to have a linear relationship with temperature difference, regulated by a film coefficient h. Radiation transfer is based on a theoretical relationship between two parallel plates, emitted flux being evaluated as the product of Stefan Boltzmann's constant (σ) a specific emissivity (ε) and the fourth power of the temperature (T). Net flux is determined as the difference between that produced by the fire environment (represented by the furnace temperature T_f with the total effect of the environment lumped into a single furnace emissivity ε_f) and that emitted by the surface (controlled by ε_s & T_s).

Radiative heat transfer is considered as dominant, particularly as a fire test progresses and the temperature increases greatly (considering the fourth power governing effect) but convective transfer may be significant in the initial stages, although the effect varies considerably depending on the fire environment.

The net heat flux q_n is the important thermal impact as far as structural damage is concerned. This is driven by the total flux q_f produced by the fire, the boundary elements and the hot gas layer reduced by the re-emitted flux q_s from the surface of the structure. The emitted flux itself depends greatly on the material of the enclosed structure and the nature of the surface presented towards the fire as these control the rate of heating of the surface T_s and hence its emissivity e_s. Contrarily, flux measurement devices are required for the evaluation of total flux q_f. Most devices will facilitate the direct production of a magnitude for net flux q_n and thus quantification of the emitted surface flux q_s is additionally required to assist the quantification of q_f

Heat flux measurement by Gardon gauge or steel billet is based on the heat transfer processes represented by Eqn. (1). In the Gardon gauge, the surface of the measuring device is maintained at a low temperature by water cooling, such that components in Eqn (1) due to T_s are insignificant, which simplifies calculation. In some cases, a sapphire window may also be applied to the apparatus in order to restrict energy transfer to the measurement device to radiative terms only. In the steel billet, the surface temperature may rise to significant levels, so accurate evaluation of particularly the emitted radiation term $\sigma\varepsilon_s T_s^4$ may become an important issue. In this respect pre-treatment of the surface to maintain a blackened coloration is important so that a known consistent emissivity ε_s may be assumed.

4 General programme

The current research project [3] is still in its early stages, although significant preparatory work has been completed. An instrumentation station has been designed (detailed later), which facilitates flux measurement by means of a Gardon gauge and a steel billet device. At the same location gas temperatures can be monitored using standard thermocouples and the plate thermometer. The stations are designed to be located within the floor slab of the ceiling of fire test compartments in order to assess the heat flux incident upwards on the exposed surface of the structural slab from the fire environment below.

The research is focused on using such instrumentation stations to assess fire severity in four compartment fire tests planned within the Cardington fire test programme. In the two corner tests conducted to date the predominant outputs were flux-time curves (total combined flux incident on the plane of instrumentation location) together with adjacent gas temperature-time curves. In these tests two instrumentation stations were used to cover the quite small test areas. In the two larger compartment tests to take place within the first six months of 1996 it is planned to supplement the data from the two complete instrumentation stations by including a larger number of steel billet devices spread over the larger test areas.

Data collected from the compartment fires will be compared with similar measurements taken using the same instrument stations in standard fire test furnaces, mainly at the University of Ulster. However the use of other UK test centres are currently being considered. In addition to these tests, an independent laboratory programme, evaluating the accuracy and utility of the billet device for flux measurement over a range of constant flux, fire growth and decline scenarios, is currently in progress.

The output from the overall research project will be a comparison of the fire growth from the four compartment tests utilising flux-time and temperature-time measurements, a comparison of these outputs with the output data from standard fires, and an evaluation of the steel billet as a flux measurement device. The current information available relates to data collected from the recent two corner room tests.

5 The instrumentation stations

Details of the standard instrumentation station used [13] are shown in Figure 1.

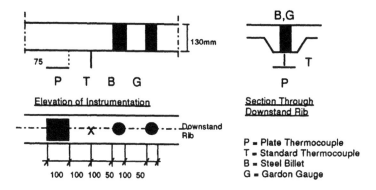

Figure 1. Details of an instrumentation station

Measurement comprised quantification of the net surface heat flux transfer incident on the plane surface of the floor slab comprising the upper surface or roof of the compartment, together with evaluation of the hot gas temperature in the proximity of the flux measurement instrumentation. The instrumentation station was aligned on the downstand rib of the metal-decking composite trough floor. A Gardon gauge and a steel billet were fitted within circular holes in the floor slab trough and located such that the surface measurement point on each device was aligned flush with the bottom surface of the trough. A standard bead thermocouple and a plate thermometer were positioned about 75 mm below the line of the trough decking and located on plan so as not to significantly restrict the 'view' of the flux instrumentation from the fire environment.

5.1 Hot Gas Temperature

Bead thermocouples and the plate thermometer were the standard temperature measurement devices. The thermocouples used were 3 mm inconel sheathed type k devices. The plate thermocouple was of standard design [1], notionally a 100 mm square 0.7 mm thick flat inconel steel plate, with a stiff insulation backing protecting the thermocouple which was braised onto the centre of the steel plate.

5.2 The Gardon Gauge

The Gardon gauge is a direct heat flux measurement device, which relies on a differential temperature measurement between the centre of a 10 mm diameter constantan face plate and a grounded reference temperature supplied by a cooled copper jacket attached circumferentially. The flux is determined using a linear calibration against the differential temperature and the average temperature is also recorded for information and to ensure that the device does not become overheated during the test.

As shown in Figure 2, the Gardon gauge was fitted into a hole cored in the floor slab by means of an adjustable extension sleeve, packed with insulation and located vertically

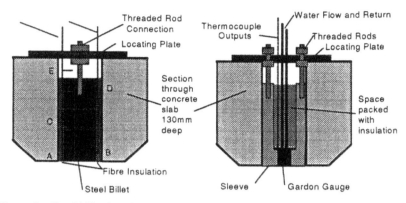

Figure 3.   Steel billet installation   Figure 2.   Gardon gauge installation

such that the exposed face plate was in line with the surface of the downstand trough. Cooling water was circulated from a header tank through plastic piping driven by an electric pump.

The disadvantages of the Gardon gauge are its initial expense, its lack of robustness as surface damage to the face plate can cause loss of calibration and the differential thermocouple is also sensitive to damage, as well as the very laborious additional requirement of the provision of a circulating water supply.   The advantage is that the single direct millivolt output data can be converted directly to a flux magnitude.

5.3 The Steel Billet

Installation details of the steel billet are shown in Figure 3.   The device consists of a 40 mm diameter x 100 mm long steel billet, thermocoupled at various positions along its length and with one face (the exposed face) heat treated to produce a consistent blackened surface. The billet was wrapped in insulation along its length, with its downstand face presented towards the fire, located so that the exposed face was flush with the underside of the floor slab. The opposite face was also left exposed to free air.   Thermocouples were embedded in pre-drilled holes within the billet at various locations longitudinally, centrally (C) and within 10 mm from each end (B,D) plus close to the exposed surface (A) and in some cases at an additional position within the air near the unexposed face.

The objective behind the design and installation of the billet was to create quasi one-dimensional heat flow longitudinally to facilitate the use of a fairly simplistic mathematical model representing heat diffusion within the element. This computer model may then be used as a black box representation of the instrumentation. The proposal is that input flux-time distributions may be tuned against the black box to produce computer simulated temperatures similar to collected test data.   A finite difference computer model provides the heat diffusion analysis and thermocouples B, C and D were located to be co-incident with the finite difference grid.   Application of this simple one dimensional finite difference analysis on a spreadsheet allows the inverse thermal analysis of the flux-time distribution without the need for complex analytical techniques. Initial laboratory trials of this methodology have proved successful.   In addition, a more simple method of providing an assessment of flux magnitudes by direct calibration from an instantaneous initial temperature gradient, gleaned from thermocouples A and B, is proving fruitful, but possible limitations of this approach are anticipated.

Two disadvantages of the billet are its lack of sophistication and its largely unproven performance. In addition, another disadvantage is that at least four thermocouple data points are required for each billet and the interpretation of the test data requires some post-processing. The advantages of the billet are its robustness, its ease and cost of manufacture and its simplicity of installation and operation.

6 Test results

Some test data from the recent two corner room tests are presented in Figures 4–7. In Test 1 [14] (July 95) the compartment floor area was some 80 m^2, having a wood crib fire load density of 45 kg/m^2, with a controlled ventilation that promotes a steady growth and decay phase. A single instrumentation station was installed at an internal location distant from the ventilation opening. In Test 2 [15] (Oct. 95) the compartment area was smaller 54 m^2, fuelled by wood cribs of fire load density 40 kg/m^2. In this case, the opening areas were glazed and initially fire growth was retarded, however natural fire growth and decay phase ensued after part of the glazing area was broken. Two instrumentation stations were installed in this case, one quite close to the window ventilation openings (No. 1) and the other internally (No. 2) in a similar position to the previous test.

Figure 4 shows the fire growth of Test 1 in gas temperature terms, temperature rising steadily to approximately 1000 °C before decaying. An interesting feature is the plate thermometer response. The device lagged in time and temperature response, during the growth stage denoting a different thermal mass and inertia from that of the thermocouples. In the decay phase, however, the temperature drop of the plate thermometer lagged the thermocouple, which being a 1.5 mm sheathed type exhibited rapid response to temperature fluctuation.

Figure 5 shows gas temperature data for Test 2. The initial suppressed growth is evident and both inner and outer stations show quite similar temperature outputs (TC1 and TC2). The plate thermometer (PT2) shows a temperature lag but no time lag. In this test the bead thermocouples were larger, 3 mm sheathed. Figure 6 shows the flux-time variations measured by the Gardon gauges at the two stations. Flux-time distributions followed the temperature profiles in tracking fire growth and decay but ,notably, the internal station flux distribution (GG2) was much more severe than that of the outer station (GG1), even though the measured temperatures of Figure 5 were similar. This highlights an inadequacy of temperature being a measure of fire severity. Finally, Figure 7 gives a comparison of Gardon gauge fluxes and billet derived fluxes. At this stage only a crude comparison of distributions has been made and observations must be considered preliminary. Estimates of billet flux magnitudes have been based on the temperature difference between the thermocouples A and B, adjacent to the billet exposed surface, and scaled on the secondary Y axis to infer a comparison. Some time lags may be observed and anomalies are evident in the peak values of station 2. However, general comparisons between devices are evident and the gross differences in fluxes measured between stations by the Gardon gauges were reinforced by the billet responses.

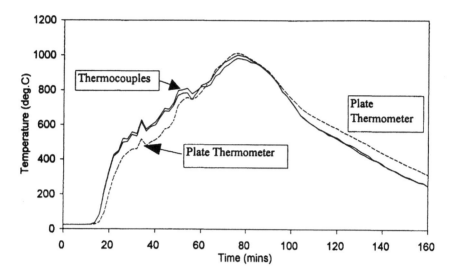

Figure 4.   Test 1: Compartment gas temperature data

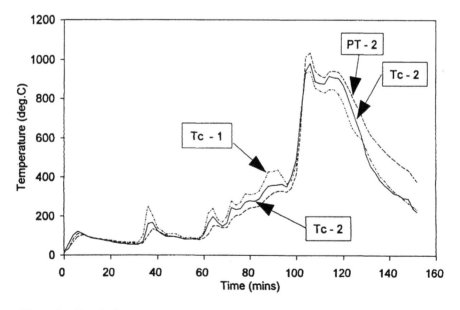

Figure 5.   Test 2: Compartment gas temperature data

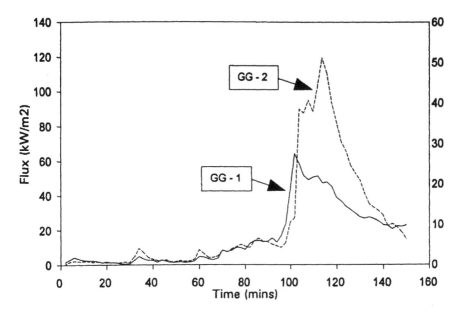

Figure 6.   Test 2. Gardon gauge flux measurements

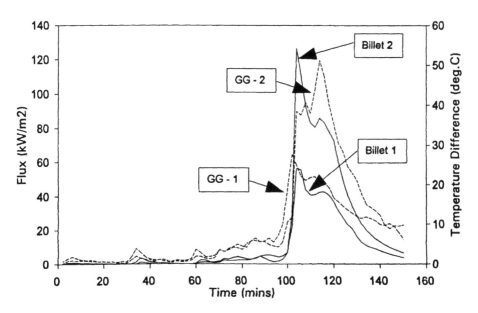

Figure 7.   Test 2: Comparison of flux measurement devices

7 Conclusion

Although the overall project is still in its early stages, initial experiences have been positive. The design of the instrumentation stations has been refined to permit ease of installation and utilisation. Data from the two initial corner tests have provided interesting output data which must be examined in more detail. Initial billet derived flux distributions from the corner compartment tests give some indication of the utility as a means of flux assessment. This premise is supported by additional calibration work presently being carried out in the associated laboratory programme.

8 References

1.   Wickström, U., 'The Plate Thermometer - *A Simple Instrument for Reaching Harmonised Fire Resistance Tests*, NORDTEST Project 609-86, SP Report 1989:03, Swedish National Testing Institute, Fire Technology, Borås 1988. 43pp.

2.   O'Connor D.J., Silcock G.W.H. & Shields T.J., *An Investigation of a Model Testing Strategy for Thermo-structural Performance Under Standard Fire Regimes*, EPSRC Grant No. GR/H 44455, Oct 1992 - Oct 1995.

3.   O'Connor D.J., Silcock G.W.H. & Shields T.J., *Heat Flux Measurement in Real and Standard Fires*, EPSRC Grant No. GR/K 75743, Oct 1995 - Oct 1996.

4.   Sterner, E., Wickström, U., *Calibration of Fire Resistance Test Furnaces*, NORDTEST Project 785-88, Swedish National Testing Institute, Borås 1989.

5.   Cooke, G., *Can Harmonisation of Fire Resistance Furnaces be Achieved by Plate Thermometer Control?*, Building Research Establishment Document PD292/93, 12pp.

6.   Cooke, G., *The Severity of Fire in a Large Compartment with Restricted Ventilation*, Building Research Establishment Document PD331/93, 9pp.

7.   Kay, T.R., Kirby, B.R. & Preston, R.R., *Calculation of the Heating Rate of Unprotected Steel Member in a Standard Fire Resistance Test*, British Steel Technical, Swinden Laboratories, (submitted to Fire Safety Journal), 1995, 13pp.

8.   Magnusson, S.E., Thelandersson, S., (Harmathy, T.Z.,), *A Discussion of Compartment Fires.*, Fire Technology, 10 (3), pp 228-246.

9.   Paulsen, O.R., Hadvig, Sven, *Heat Transfer in Fire Test Furnaces.*, Journal of Fire and Flammability, 8, pp 423-442.

10.   Babrauskas, V., Williamson, R.B., *Temperature Measurement in Fire Test Furnaces*, Fire Technology, 14, pp 226-238.

11.   Sultan, M.A., Harmathy, T.Z., Mehaffey, J.R., *Heat Transmission in Fire Test Furnaces*, Fire and Materials, 5 (3), pp 112-123.

12.   Blackwell, B.F., Douglass, R.W., Wolf, H., *A User's Manual for the Sandia One-Dimensional Direct and Inverse Thermal (SODDIT) Code*, Sandia National Laboratories, 1987.

13.   O'Connor, D.J. & Dowling, J.J., *Specification for Heat Flux Instrumentation at Cardington Fire Tests*, University of Ulster Research Report No. BE/4122-D0, Mar. 1995, 9 pp.

14.   O'Connor, D.J. & Morris, B., *Cardington Fire Test: Corner Room Test: July 1995: Heat Flux Test Results*, University of Ulster Research Report No. BE/4122-D1, Dec. 1995, 11 pp.

15.   O'Connor, D.J. & Morris, B., *Cardington Fire Test: Corner Room Test: October 1995: Heat Flux Test Results*, University of Ulster Research Report No. BE/4122-D2, Jan. 1996, 12 pp.

LARGE COMPARTMENT FIRE TESTS

T. LENNON
Structural Design Division, Building Research Establishment, UK

1 Introduction

There is a growing opinion that the structural contribution of composite steel deck floor systems is under-utilised in current design procedures, particularly for the fire limit state. This, together with evidence from three-dimensional numerical models and investigations from real fires such as Broadgate that the fire resistance of complete structures is significantly better than that of the single elements from which fire resistance is universally assessed, has led to a demand for full scale fire tests.

Computer programs have developed way beyond available experimental data, however, before such analytical techniques can be used with any confidence it is necessary to verify them against test results from real buildings subject to real fires. Whilst it is possible to study structural behaviour by examining fire-damaged buildings, interpretation of the findings is complicated by the lack of information on heating rates, temperatures and the stresses imposed on the members at the time of the fire.

The concept of equivalent time of fire exposure, which relates the duration of heating in a standard test furnace to the thermal loading received in a real post-flashover fire, has recently been introduced into the fire part of Eurocode 1. However, this approach has only been validated on compartments with dimensions significantly smaller than found in modern office buildings. Large compartment fire tests in real buildings would complement work recently carried out by the Fire Research Station (FRS) on natural fires in small to medium compartments. It is therefore desirable to undertake a series of full-scale large compartment fire tests in real structures to improve the design procedures for modern steel-framed buildings and to quantify safety margins.

The development of the Building Research Establishment's (BRE) Large Building Test Facility (LBTF) at Cardington has given researchers a unique opportunity to carry out a series of controlled fire tests on a building designed and built to current practice guidelines.

A series of fire tests are being undertaken on a purpose built eight-storey, three bay by five bay steel-framed building with composite floors and overall dimensions 33.5 x 21 x 45 m at the Cardington LBTF, near Bedford, England. As part of the overall fire programme, BRE have carried out a fire test on a compartment (9 m by 6 m) representative of a corner office, and in April 1996, a test will take place on a large compartment (21 m by 18 m) representative of a large open plan office. The fuel source in both cases is timber cribs which give a fire load of 40 kg/m^2. An applied loading of 5.48 KN/m^2 representing the dead load plus one third of the imposed loading will be present on all floors. The response of the structure will be measured using a range of instrumentation – the thermal response will be recorded using both thermocouples and heat flux transducers whilst the structural response will be monitored using strain gauges, displacement transducers and rotation transducers. In addition a video recording of each of the tests will be made. The influence of glazing, (a highly unpredictable parameter) on the development of the fire will also be investigated.

The objectives of the programme are to examine the behaviour of multi-storey steel-framed buildings subject to real fires and to use the data from the tests to validate computer models for structural analysis at elevated temperatures, and assess the fire parts of the forthcoming Eurocodes. The work will provide substantial benefits and produce high quality data which will aid decisions on the degree of fire protection required for steel-framed buildings, leading to reduced costs though maintaining existing levels of safety. This paper describes the philosophy and objectives of the programme with particular reference to the corner fire test. The type and location of the instruments used to measure the response of the building to the corner fire are enumerated. Preliminary results from the corner fire are presented and discussed and mention is made on the nature of the forthcoming large compartment fire test.

2   Corner fire test

The BRE corner fire test took place on the evening of 23rd October 1995. The fire was ignited on the second floor of the eight-storey building in a corner compartment bounded by gridlines E to F and 3 to 4 (figure 1). Figure 2 is a three-dimensional view of the compartment. For ease of reference the main steel members within the compartment are identified according to the schematic shown in figure 3. This was extended to identify members adjacent to the compartment as shown in figure 4.

2.1 Design of the compartment

2.1.1 Location

The choice of the third floor (that is fire on the second floor testing the third floor) as the location for the fire test was agreed in order to minimise the heat rise on the structural members of the hanger itself, and to facilitate observations during the test. The first and second floors were not typical of the building due to the presence of a central concourse.

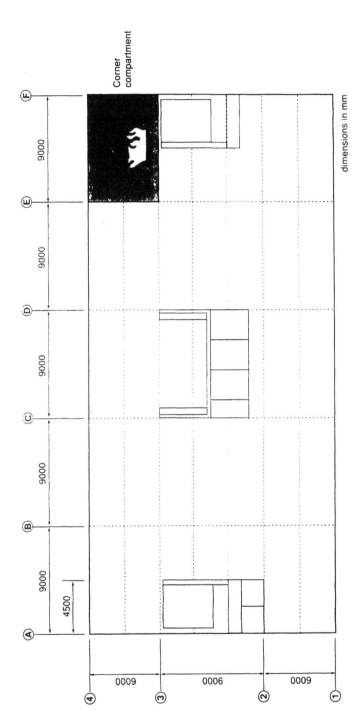

Corner compartment

dimensions in mm

Figure 1.   Proposed corner compartment test

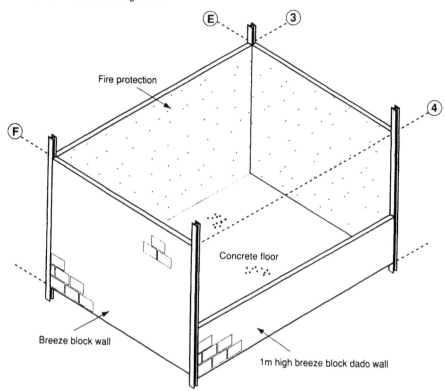

Figure 2.   3-D view of corner compartment

The location was also dictated by the need to ensure that, as far as possible, the results of a test in one location do not adversely affect subsequent tests. This required a degree of co-ordination between all the parties involved in the fire test programme, principally BRE and British Steel Technical.

2.1.2 Protection

The compartment itself was created using fire resistant board running between the columns forming the boundaries of the compartment. The internal column on gridline E3 was fully protected. Protection was also provided to the two external columns E4 and F4. The two remaining columns were outside the compartment, behind the shaft walling used to protect the stairwell at the Eastern end of the building. The secondary beam running through the centre of the compartment was unprotected. The presence of the glazing formed the only protection to the edge beam running between gridlines E4 to F4.

2.1.3 Compartmentation

As mentioned above the compartment was bounded on the Southern end by a fire resistant partition extending from the shaft walling to the column on gridline E3.

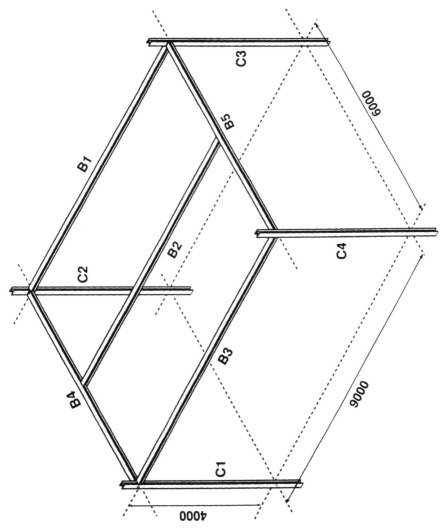

Figure 3.   Location of beams and columns

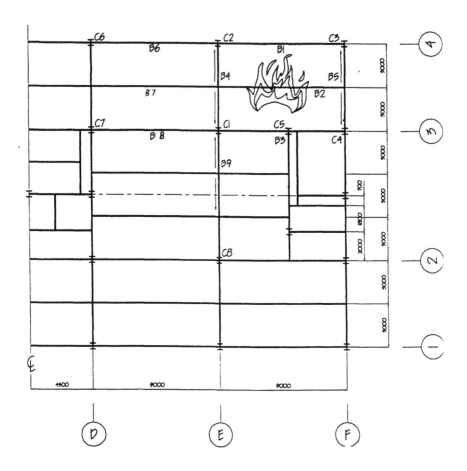

Figure 4.   Plan of 2nd – 3rd floor steel

The Western boundary was similarly formed by constructing a fire resistant partition between columns E3 and E4. The Northern boundary was formed by constructing glazed aluminium screens between columns E4 and F4, while the Eastern boundary was the gable wall already in existence. No attempt was made to provide any additional restraint to the wall.

2.1.4 Ventilation

No additional ventilation was provided and no attempt was made to artificially seal the compartment. One of the purposes of the test was to investigate the influence of the glazing in the development of the fire. For the purposes of design it is assumed that all the glass has cracked at the outset and that the amount of ventilation present is the maximum possible. Results from the test would suggest that this is an over-simplification.

2.2 Loading

2.2.1 Thermal loading

Twelve timber cribs were placed in the compartment. The fire load was 40 kg/m^2 over a floor area of 54 m^2 giving a total fire load of 2,160 kg. Each crib was constructed from 200 sticks of 50 x 50 mm rough sawn softwood with ten sticks in each of 20 layers giving a crib height of 1 metre. Ignition was from a single point close to the stairwell.

2.2.2 Static loading

Floor loads of 5.48 KN/m^2 were provided by sandbags to represent the dead load plus one third of the imposed loading on all floors, other than the second floor where the bags were split and the sand spread uniformly over the floor area. The bags on the floor above were fitted with slings attached to the steelwork which would support the load should the displacement exceed a pre-determined value of 400 mm.

2.2.3 Instrumentation

The primary requirements of the instrumentation were to measure the temperature of the structural and non-structural components in the fire compartment, and of those components immediately outside the compartment; to measure the distribution of internal forces in the components adjacent to the compartment; to measure the deflected shape of the floor, end-wall and main structural members and to measure the heat transfer through the floor and walls of the compartment. These requirements were met as detailed below:

- *Temperature.* In total 278 thermocouples were used to monitor the temperature of the steel columns and beams within the compartment, the temperature through the concrete slab and the atmosphere temperature within the compartment. Additional instruments were used to monitor the temperature immediately outside the compartment, the temperature of the hanger steelwork, indicative temperatures of unprotected steel sections suspended from the ceiling of the compartment and the internal and external temperatures of the gable wall.
- *Strain.* A total of 300 strain gauges were used to measure the response of the structure to the fire. Eight columns on the fire floor, the floor below, the floor above and the seventh floor were instrumented in addition to four beams on the fire floor. Additional strain gauges were used to measure the response of the reinforcing mesh and the concrete surface on the third floor.

- *Displacement.*   23 no. 1000 mm travel displacement transducers were used to measure the deformation of the concrete slab on the third floor. An additional 24 no. 100 mm travel transducers were used to measure the axial and lateral movement of columns E3 and E4. Twelve clinometers were used to measure the major axis rotations of the connections within the compartment while an innovative laser system monitored the movement of the gable wall.
- *Miscellaneous.*   In addition to the instrumentation mentioned above, the University of Ulster monitored heat transfer through the floor while the Fire Research Station measured heat transfer in the plane of the window.

3 Preliminary results

3.1 Atmosphere temperature

The development of the fire was largely influenced by the lack of oxygen within the compartment. After an initial rise in temperature the fire died down and continued to smoulder until the fire brigade intervened to vent the compartment. The removal of a single pane of glazing resulted in a small increase in temperature followed by a decrease. Flashover did not occur until a second pane, immediately below the first, was removed. This initiated a sharp rise in temperature which continued as the fire developed. The maximum recorded atmosphere temperature in the centre of the compartment was 1051 °C 1200 mm from the ceiling after 102 minutes. The maximum atmosphere temperature recorded was 1059.5 °C to the South of the compartment 1500 mm below the ceiling.

3.2 Steel temperatures

The maximum recorded steel temperature of 903 °C occurred after 114 minutes on the bottom flange of the unprotected beam B2 in the middle of the section. The maximum temperature reached by the edge beam was 690 °C, again after 114 minutes, despite being completely engulfed in flames. The maximum temperature of the beam framing into the stairwell was 629 °C on the web of the section at mid-length.

3.3 Displacement

The maximum recorded value of slab displacement occurred in the centre of the slab after 130 minutes where it reached a value of 269.4 mm, however by the next morning the slab had recovered to a final displaced position of 159.7 mm.

4 Discussion

Initial observations have highlighted a number of areas worthy of consideration. The influence of glazing on the development of the fire requires further study, and it is hoped that any glazing present in the large compartment test will be instrumented. The masonry wall forming the Eastern boundary of the compartment retained it's integrity despite a significant thermal gradient across the wall and substantial lateral deformation. The fire resistant partitions performed adequately and prevented any appreciable heat rise outside the boundaries of the compartment.

There was significant lateral-torsional deformation of the secondary beam running through the shaft walling, whilst the primary beam forming the Western boundary of the compartment remained virtually straight. While this may in part be due to the position of the partitions relative to the underside of the lower flange of the beams, it is more likely that the enhanced performance of the primary beam was due to the restraint provided by the secondary beam framing into the web of the member halfway along its length.

That any damage was limited to the area within the compartment demonstrates the integrity of the structure. An initial comparison with analytical techniques suggests that a simplified theory of tensile membrane action provides a quick and accurate method of predicting the behaviour of the composite slab subject to a fire.

In total some 1000 channels of data were recorded every 2 minutes for the duration of the test and at 10 minute intervals overnight. A great deal of detailed analysis remains to be done although initial studies suggest that the test has provided a large store of valuable data which, given time, can be used to inform the fire resistant design procedure of steel-framed buildings with composite floors.

5 Large compartment fire test

In order to provide further verification for analytical methods of predicting structural behaviour at elevated temperature, the BRE will be conducting a large compartment fire test on the eight-storey building in April 1996. The fire will take place on the same floor as the corner test but on the Western end of the building.

As with the corner fire test the fuel source will again be timber cribs with a fire load of 40 kg/m^2. The main difference between the two tests will be one of scale. The compartment will be formed by constructing a fire resistant wall across the width of the building in front of gridline C., and the total area of the compartment will be 342 m^2.

TENSILE MEMBRANE ACTION IN SLABS AND ITS APPLICATION TO THE CARDINGTON FIRE TESTS

Y.C. WANG
Structural Design Division, Building Research Establishment, UK

Abstract

Most UK building codes and standards adopt the philosophy of limit states for structural design. For a structural member, one of the most important limit state is its ultimate strength. For a reinforced concrete slab, this ultimate strength is usually calculated assuming that failure occurs by pure flexural bending, using the yield line theory to give an upper bound solution. However, various experimental and analytical studies indicate that the effect of membrane actions can increase the ultimate load of the reinforced concrete slab.

This paper addresses the problem of tensile membrane action in reinforced rectangular concrete slabs. Park's equation for rectangular slabs with restrained edges is given first. Kemp's treatment of membrane action in a simply supported isotropic square concrete slab is then introduced. The main body is devoted to a more detailed description of tensile membrane action at large deflection in a rectangular reinforced concrete slab with various partial-strength support conditions. These different theories are then applied to the results from the fire tests on the Cardington steel-framed building to examine their validity and to predict the outcome of future fire tests.

1 Introduction

When part of a building is involved in a fire, temperatures in the surrounding structural members increase. Since material strength and stiffness reduce at elevated temperatures, the load carrying capacities of these members also decrease. The failure limit state of a structural member is reached if the load carrying capacity of the member is reduced to such an extent that it is lower than the load applied to the member.

At present, this limit state is not allowed. Therefore, current design guidance recommends that a structural member should be protected to reduce its maximum temperature rise in the event of a fire so that its load carrying capacity is higher than its applied load at the specified maximum temperature.

Steel has a very high coefficient of conductivity, and if exposed to fire its temperature rises very rapidly. According to the present fire resistance requirement, structural steel members should be protected for a minimum fire resistance period.

Whilst the principle of fire protection to structural members to prevent structural collapse in the event of a fire is sensible and the fire protection practice has so far been safe, its applicability in the context of the fire resistance of a complete building should be critically examined for the following reasons:

1. The current approach to assess the fire resistance of a structural member is based on either a limited deflection in a standard fire resistance test or a structural failure at small deflection [1]. Since the main objective of structural fire safety design is to prevent total collapse of the structure, large deflection in structural members can be tolerated. Hence deflection control ceases to be meaningful. More importantly, a structural member may be stable at large deflections after the first failure mode at small deflection.
2. The fire resistance of a complete structure is an entirely different concept to the fire resistance of a single structural member. Whilst the failure of an individual member in fire means that it is no longer safe to sustain the applied load, the failure of some structural members in a fire does not necessarily endanger the safety of the complete building. Sometimes, the fire resistance of the complete building may be independent of these failed structural members because of the ability of the rest of the structure to develop an alternative load path to bridge over these failed structural members.

In a modern steel framed composite building with reinforced concrete floor slabs, membrane action has been identified as the primary mechanism to account for the stability of a structural member at large deflection and for providing an alternative load path after the failure of some structural members. The objectives of this research are threefold:

1. to establish a theory to describe membrane action in a rectangular reinforced concrete slab
2. to use this theory to analyse the behaviour of the Cardington fire tests
3. to demonstrate the potential of this mechanism as a means to achieving complete elimination of fire protection to some steel members and preventing the progressive collapse of buildings subject to other types of accidental loading.

2 Membrane actions in a reinforced concrete slab

2.1 General behaviour

Figure 1 illustrates schematically the load-deflection curves of reinforced concrete slabs with clamped and simply supported edges.

Calculations for the load carrying capacity of a reinforced concrete slab are based on the slab failing under pure flexural bending. This represents the first failure mode of a slab at small deflections. Johansen's [2] yield line theory is often thought to give an upper bound (point A in figure 1) to the load carrying capacity of the slab. This theory has been adopted in many codes of practice for the design of reinforced concrete slabs.

However, if the slab is laterally restrained, the slab will arch from boundary to boundary. This results in the development of a compressive membrane force in the concrete slab, increasing its load carrying capacity.

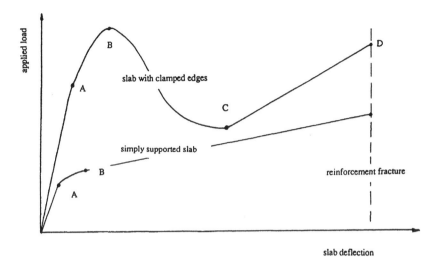

Figure 1.   Complete load-deflection curve for a reinforced concrete slab

Tests on buildings [3] and laterally restrained individual slabs [4] confirmed the enhanced load carrying capacity of concrete slabs due to compressive membrane action. The load carrying capacity of the slab at maximum compressive force (point B in figure 1) may be several times the strength according to yield line theory.

As the slab deforms further, the depth of the cracked concrete increases and the available uncracked concrete for compressive stress diminishes. The membrane action in the slab changes from compressive to tensile. When cracking extends over the entire depth of the concrete cross-section (point C in figure 1), the applied load on the reinforced concrete slab can be regarded as being taken by the tensile membrane action of the steel reinforcement. During this stage of loading, the applied load on the slab is supported by the net of reinforcement as illustrated in figure 2. The load carrying capacity of the slab increases with increasing slab deflection. The slab collapses when the slab reinforcement fractures at point D in figure 1.

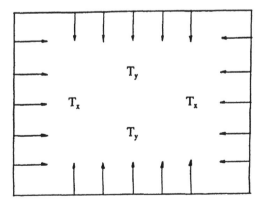

Figure 2.   Tensile membrane load-carrying mechanism in a slab with clamped   edges

For a simply supported slab, its ability to develop compressive membrane action is very limited. However, at large deflection, a slab can develop an in-plane ring beam in compression to support the development of tensile membrane action in the central region of the slab. The load carrying mechanism in a simply supported slab is drawn in figure 3. This behaviour was confirmed by Brotchie and Holley's tests [5]. For a simply supported slab, there is a smooth transition from pure flexural behaviour at small deflection to tensile membrane action at large deflection as shown in figure 1.

Figure 1 clearly shows that for a reinforced concrete slab with clamped edges, although compressive membrane action may be used to improve the load carrying capacity of a slab in conventional design at small deflection, its behaviour is unstable. Furthermore, compressive membrane behaviour is extremely sensitive to edge restraints and initial imperfection. These are the obstacle to the wider acceptance of compressive membrane action by designers.

Tensile membrane action is stable, but it occurs at very large deflections in reinforced concrete slabs. For normal design at the cold condition, these large deflections would violate serviceability conditions. However, deflection ceases to be a problem when the slab is subjected to an accidental loading such as a fire. It is therefore acceptable to explore the enhanced load carrying capacity of a reinforced concrete slab due to tensile membrane action at large deflection to assess its collapse strength for fire safety design.

The effect of tensile membrane action in a reinforced concrete slab attracted some attention in the late fifties to early seventies. Park [6] derived an equation for a rectangular slab with clamped edges. Wood [4] presented a unique method to calculate the effect of tensile membrane action in a simply supported isotropic circular slab and Kemp [7] followed this approach to find the solution for a simply supported isotropic square slab.

Recently, the finite element method has been used to determine the load-deformation relationship of a concrete slab, including both compressive membrane action and tensile membrane action using the displacement control method[8, 9]. However, this method is rather complex to use.

In this paper, a methodology is developed to evaluate the load-deflection curve of a reinforced concrete slab with various partial-strength support conditions at large deflections.

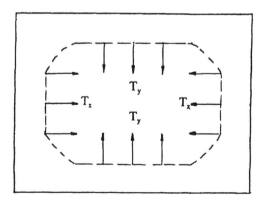

Figure 3.   Tensile membrane load-carrying mechanism in a simply supported slab

2.2 Tensile membrane action in a laterally restrained rectangular slab

Using the standard membrane theory, Park [6] derived the following relationship between the uniform load density on a slab and its central deflection under the effect of pure plastic tensile membrane:

$$\frac{pL_y^2}{T_y w_0} = \frac{\pi^3}{4\sum_{n=1,3,5\ldots}^{\infty} \frac{1}{n^3}(-1)^{(n-1)/2}[1-\frac{1}{\cosh[(n\pi L_x/2L_y)\sqrt{T_y/T_x}]}]} \tag{1}$$

Where L_x and L_y are the length of the longer and shorter spans of the slab; T_x and T_y are the tensile forces of the reinforcement in the longer and shorter spans respectively. w_0 is the central deflection of the slab at fracture of the reinforcement (point D in figure 1).

Park [6] proposed to use a deflection of 10% of the short span, i.e. $w_0 = 0.1L_y$ at reinforcement fracture. Black [10] found that both equation (1) and a deflection to span ratio of 0.1 underestimate the tensile membrane effect. This is because equation (1) was developed on the assumption of pure tensile membrane action, while additional loadbearing capacity due to uncracked concrete exists in realistic slabs, especially in deep concrete slabs. At total collapse, the deflection was found to be of the order of 15% of the short span. Therefore, equation (1) together with a deflection/span ratio of 0.10 gives a safe estimate of the collapse load of a reinforced concrete slab.

2.3 Tensile membrane action in a simply supported square slab

Following the approach of Wood [4] for a circular slab, Kemp [7] derived equations for the compressive membrane action and the tensile membrane action in a simply supported square slab with isotropic reinforcement. His analysis adopted a diagonal yield line mechanism for the square plate. This was justified on the ground of symmetry.

Kemp's [7] analysis indicated that there is limited strength enhancement due to compressive membrane action in the slab at small deflection. This strength enhancement factor is:

$$\frac{p}{p_j} = 1 + \frac{w_0^2}{64\beta}(\frac{T_0}{M_0})^2 \tag{2}$$

p   the ultimate applied load,
p_j Johanson's load according to yield line theory,
w_0 the slab central deflection,
β = $0.75\chi/(1-0.75\chi)$, in which χ is the reinforcement ratio
T_0 reinforcement yield load per unit concrete width
M_0 slab unit moment capacity when there is no membrane force

The condition is that the central deflection w_0 is less than a critical deflection w'_0 which is the slab central deflection at which the concrete starts to crack throughout its depth. This critical deflection is given by the equation $w'_0 = 8\beta M_0/T_0$. Once the slab starts to crack throughout its depth, tensile membrane action takes over as the load carrying mechanism. The strength enhancement factor for slabs under tensile membrane action is:

$$\frac{p}{p_j} = 1 + \beta[\sqrt{\frac{w_0 T_0}{2\beta M_0}} - 1]^2 \tag{3}$$

If the same diagonal yield line pattern is adopted for rectangular slabs with isotropic reinforcement, Kemp's equations (2) and (3) are directly applicable.

2.4 Tensile membrane action in a rectangular slab with partial strength supports

Both Park [6] and Kemp's [7] theories of tensile membrane action apply to particular examples of a rectangular slab with specific partial strength edges. A literature survey revealed that no general solution is available to evaluate the load-deflection behaviour of a rectangular slab with partial strength edge supports. However, Kemp's [7] general procedure can be extended to such a slab.

Space is limited for the inclusion of the detailed derivation. Therefore, only the assumptions, the types of slab considered, the procedures of derivation and a few key equations are given in this paper.

2.4.1 Assumptions

The main assumptions used in this study are:

- The slab behaves in a rigid-plastic way. Yield lines in the slab form at zero deflection.
- These yield lines remain unchanged during further slab deflection.
- Deflections in the edge supports are very small compared to those in the central slab.
- Full edge restraining force capacities in the slab plane develop at zero deflection. The edge restraining force capacity is the least of the edge support lateral load carrying capacity and the slab reinforcement capacity.
- Edge movement in the slab plane is uniform along the edge length.

2.4.2 Type of edge support conditions

Figure 4a shows three types of slab considered in this study. Due to the fact that the slab should be in overall equilibrium in both X and Y directions, these three types represent all possible arrangements of a rectangular slab with partial strength edges. Figure 4b gives the yield line patterns adopted in this study. The forces acting on each block bound by yield lines are shown in figure 4c. Figure 4d illustrates the various deformations along yield lines.

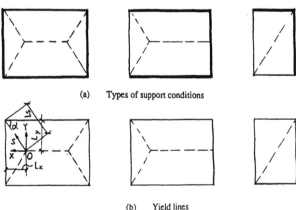

(a)   Types of support conditions

(b)   Yield lines

Figure 4a/b.   Input data for the investigation

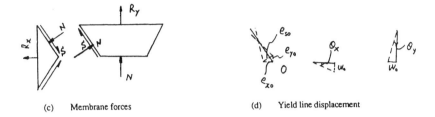

(c)   Membrane forces   (d)   Yield line displacement

Figure 4c/d.   Input data for the investigation

2.4.3 Procedures of derivation

The 5 main steps in the derivation and the key equations in each step are:

Yield condition
Kemp [7] gave the membrane force density (N, positive in compression) bending moment density (M) relationship (if the slab is not cracked throughout its depth) as:

$$\frac{N}{T_0}=\frac{\alpha}{2\beta}-\frac{\mu}{2\beta}\frac{T_0}{M_0} \ and \ \frac{M}{M_0}=1+\alpha(\frac{N}{T_0})-\beta(\frac{N}{T_0})^2 \tag{4}$$

The yield criteria (when the slab is cracked throughout its depth) change to:

$$\frac{N}{T_0}=-1 \ and \ \frac{M}{M_0}=1-\alpha-\beta \tag{5}$$

The moment is taken about the mid-plane of the concrete slab.

α and β are functions of reinforcement ratio. The parameter m is the position of neutral axis above the slab mid-plane. Its value changes along yield lines. This value can be related to the plastic neutral axis position at the centre of the slab O where yield lines intersect through the geometrical consideration.

Geometrical consideration (compatibility)
Figure 4 is referred to for the following equations. Assuming the opening along the yield line e, is perpendicular to the yield line and the position of plastic neutral axis m is obtained by dividing the yield line opening by the rotation of the yield line, it is possible

to obtain the opening of the yield line e, the plastic neutral axis m and the rotation of the yield line as:

$$e_s = \frac{e_x}{\sin\alpha} = \frac{e_y}{\cos\alpha}, \quad \mu = \mu_0 - \frac{w_0}{2L_s}s, \quad \theta_s = \frac{w_0}{L_s\sin\alpha\cos\alpha} \tag{6}$$

The values e_x, e_y and μ at the yield line intersection point O are e_{x0}, e_{y0} and μ_0. Their relationships are expressed as:

$$\mu_0 = \frac{e_{s0}}{\theta_s} = \frac{e_{x0}}{\theta_x} = \frac{e_{y0}}{\theta_y} \tag{7}$$

in which θ_x and θ_y are rotation about the Y and X axis respectively:

$$\theta_x = \frac{w_0}{L_x} = \frac{w_0}{L_s\cos\alpha}, \quad \theta_y = \frac{w_0}{L_y} = \frac{w_0}{L_s\sin\alpha} \tag{8}$$

The relationship between the lateral movement Δ_x and Δ_y may be given by the following equations:

$$\Delta_x = \frac{w_0^2}{2L_x} - e_{x0} \quad and \quad \frac{\Delta_x}{\Delta_y} = \tan\alpha \tag{9}$$

Membrane forces
The density of the membrane force in the slab can be expressed as a function of the position of the plastic neutral axis at the slab centre μ_0 by substitution of equation (6) into equation (4):

$$\frac{N}{T_0} = (\frac{\alpha}{2\beta} - \frac{\mu_0}{2\beta}\cdot\frac{T_0}{M_0}) + \frac{w_0}{4\beta L_s}\frac{T_0}{M_0}s \tag{10}$$

Substituting the appropriate values of α, β, T_0 and M_0 in the X and Y direction in equation (9) will give the membrane forces in these directions.

The yield moment can be obtained as a function of μ_0 by substituting equation (9) into equation (3).

Equilibrium condition
There are two unknowns in the slab system: the position of the plastic neutral axis μ_0 and the shear force S in the slab along the yield lines as shown in figure 4. To find the solution for these two variables, two equations are required. These two equations are in the form of two equilibrium conditions. Although four equilibrium equations can be written for the system: two for the triangular block and two for the 4 sided block, there are only two independent equations: the equilibrium equation of the triangular in the Y direction and that of the 4-sided block in the X direction. The two system variables are

found from these two equations and the value μ_0 is substituted into the appropriate equations for the calculation of the membrane force density N and plastic moment M.

Work method
Two methods may be used to obtain the load-deflection curve of the slab: the work method and the nodal force equilibrium method [4]. With the values of the edge restraining forces (R_x and R_y), the membrane forces (N) and the plastic moments (M) and their respective displacements known (Δ_x, Δ_y, e_s and θ_s), the use of the work method is more straightforward. The work method has the following general form:

$$\delta W_{ext} = \delta W_{int} \qquad\qquad (11)$$

where δW_{ext} is the increment in the external work done due to a small increase in the deflection w_0 and δW_{int} the responding increase in the internal work. The general expression for the work increment is:

$$\delta W = F.\delta u \qquad\qquad (12)$$

where F is the force and du the corresponding deflection increment due to increase in w_0.

3 Behaviour of the floor slab in the BRE corner fire test

A fire test in one corner of the Cardington steel-framed building was performed on 23 October 1995. A detailed description of this test and the fire behaviour is given by Lennon [11]. This paper concentrates on the structural behaviour of the floor slab only.

The dead load on the slab was 2.5 kN/m^2, the imposed load (sandbags) was 2.4 kN/m^2. The reinforcement was 142 mm^2/m at 50 mm below the top concrete surface. Assuming yield lines bisecting the slab corners ($\alpha=45°$) and the centre, the yield line solution is 1.95 kN/m^2. The ratio of total load (4.9 kN/m^2) to the yield line solution is 2.51. Figure 5 shows the load-deflection curve of the floor slab under tensile membrane action using the theories of Park [6], Kemp [7] and the present investigation. The corresponding deflection to the applied load according to the present investigation is 164 mm.

The slab's maximum deflection at its centre is plotted against temperature differences in figures 6 and 7. The deflection is plotted against temperature difference in the steel supporting beam in figure 6 and against concrete temperature difference in figure 7. The selection of these temperature differences as reference axes in figures 6 and 7 is because that the slab deflection is most closely related to thermal bowing.

From figures 6 and 7, it is noticed that they are two distinct phases in the deflection history of the floor slab. The demarcation occurred at the time (102 minutes into the fire test) when the fire and the steel temperatures were at their highest values. The fire temperature was over 1000 °C and the maximum steel temperature was close to the fire temperature at just below 1000 °C. The temperature difference in the steel section was also at its largest.

The following equation helps to explain the behaviour of the fire floor slab.

$$\Delta\delta_{floor} = \Delta\delta_{\Delta T(steel)} + \Delta\delta_{\Delta T(concrete)} + \Delta\delta_{mechanic} \qquad\qquad (13)$$

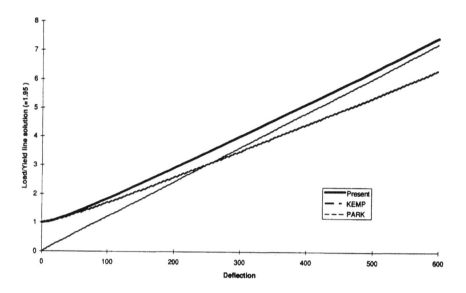

Figure 5.   Load-deflection curve, BRE corner test

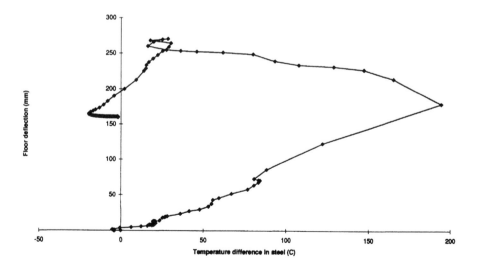

Figure 6.   Slab deflection–Steel temperature difference relationship

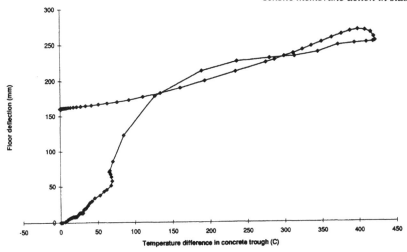

Figure 7.   Slab deflection–Concrete trough temperature difference relationship

In this equation, the total floor deflection ($\Delta\delta_{floor}$) is composed of three parts: the thermal bowing induced deflection due to temperature difference in steel ($\Delta\delta_{\Delta(steel)}$), the thermal bowing induced deflection due to temperature difference in concrete ($\Delta\delta_{\Delta(concrete)}$) and the increased deflection due to strength reduction in concrete and steel ($\Delta\delta_{mechanic}$).

Until 102 minutes, the temperature difference in concrete was very low (figure 7). Total deflection in the slab is mainly a result of temperature difference in the steel section and increased mechanic deflection. At 102 minutes, the mechanical deflection reached its maximum when both the steel beam and steel decking lost their strength at temperatures of about 1000 °C. The load on the floor slab was then resisted by the tensile membrane action in the concrete slab.

Since the applied load on the floor slab was unchanged during the fire test and the reinforcement retained its strength at temperatures lower than 300 °C, the maximum mechanical deflection stayed unchanged during the fire test after 102 minutes.

After 102 minutes, the steel temperature was very high and the steel became very "soft", in addition the temperature difference in steel was reducing. Therefore the thermal deflection due to temperature difference in the steel decreased rapidly.

The deflection increase in the floor slab was mainly a result of the thermal bowing due to the temperature difference in the concrete slab. The steel beam was almost irrelevant to the behaviour of the floor slab after 102 minutes as shown in figure 6.

When the temperature difference in the concrete started to decrease, the deflection in the floor slab recovered. The most significant fact is that the recovery in floor slab deflection almost followed the same deflection curve of increasing temperature difference in the slab as shown in figure 7.

The final deflection in the slab was about 160 mm. This deflection is very close to the calculated deflection under tensile membrane action. This final deflection would have been reached if the fire intensity had been reduced after 102 minutes.

Since the effect of prolonged fire attack on the floor slab would be to increase the heat conduction through the concrete floor slab slowly without much detrimental influence on the strength of the reinforcement, it is possible that the floor would have survived a fire load greater than the test fire load of 40 kg wood/m². In fact, the stability of the building may be insensitive to the fire load as long as the temperature increase in the reinforcement steel does not result in a significant strength reduction.

4 Other fire tests

The above exercise was also carried out for the British Steel corner fire test. In this test, the shorter dimension was 7.5 m instead of 6 m as in the BRE corner test. The final deflection predicted using the membrane action theory is 260 mm.

The demonstration fire test will be carried out in a compartment of 18 m by 21 m. If all internal steel beams were to be unprotected, tensile membrane action theory predicts that the floor would be on the verge of collapse at a deflection of 1728 mm on the assumption that the internal columns do not change the slab deflection mode. This collapse would involve fracture of the reinforcement, the cracking of concrete in the slab in the centre and the crushing of concrete around the slab edges.

However, if only the central main beam along gridline B were protected, the maximum floor slab deflection would be much less at 513 mm. At this deflection, the maximum reinforcement strain would be small compared to its fracture strain and the maximum compressive force in the "concrete ring beam" would be only about half of its strength.

Due to the aforementioned insensitivity of the floor slab stability to the fire load, it would be safe to increase the fire load from the proposed 40 kg wood/m² representing an average office fire loading to a greater level of say 80 kg wood/m² being the upper limit of the fire loading.

5 Conclusion

Three theories of tensile membrane action in a reinforced concrete floor slab are described in this paper. They are Park's equation [6] for a laterally clamped rectangular slab, Kemp's [7] equations for a simply supported square slab and the author's extension of Kemp's theory [7] to rectangular slabs with partial strength boundary conditions.

The application of the author's extension of Kemp's theory [7] to the BRE corner fire test assuming simply supported edges seems to give an accurate prediction of the final deflection of the floor slab. The further application of this analysis to the future demonstration fire test in a much large fire compartment indicates that without fire protection to the main steel beam on grid line B, collapse would occur. If this beam were protected, the floor slab would be safe.

According to the theory of tensile membrane action, the stability of the floor slab is not sensitive to the fire load. A much higher fire load than the proposed 40 kg wood/m² can be resisted by the floor slab safely.

6 References

1.   British Standards Institution, British Standard BS476 Part 21, Fire Tests on Building Material and Structures, *Methods for Determination of the Fire*

Resistance of Loadbearing Elements of Construction, British Standards Institution, London, 1987

2. Johansen,K.W., *Yield Line Theory*, translated by Cement and Concrete Association, London, 1962

3. Ockleston,A.J., *Arching Action in Reinforced Concrete Slabs*, Struct. Eng., Vol 36, No. 6, June 1958, pp. 197-201

4. Wood,R.H., *Plastic and Elastic Design of Slabs and Plates*, Thames and Hudson, London, 1961, pp.225-261

5. Brotchie,J.F. and Holley,M.J., *Membrane Action in Slabs, Cracking, Deflection, and Ultimate Load of Concrete Slab Systems*, ACI Special Publication 30, American Concrete Institute, Detroit 1971, pp. 345-377

6. Park,R. *Ultimate Strength of Rectangular Concrete Slabs under Short-Term Uniform Loading with Edges Restrained against Lateral Movement*, Proc. Inst. Civ. Eng., Vol 28, June 164, pp.125-150

7. Kemp, K.O., *Yield of a square reinforced concrete slab on simple supports, allowing for membrane forces*, The Structural Engineer, Vol. 45, No. 7, July 1967, pp. 235-240

8. May, I.M. and Ganaba, T.H., *A full range analysis of reinforced concrete slabs using finite elements*, International Journal for Numerical Methods in Engineering, Vol. 26, 1988, pp. 973-985

9. Salami, A.T., *Equation for predicting the strength of fully-clamped two-way reinforced concrete slabs*, Proceedings of Institution of Civil Engineers, Structures & Buildings, Vol. 104, Feb. 1994, pp.101-107

10. Black, N.S., *Ultimate Strength Study of Two-Way Concrete Slabs*, Journal of Structural Division, Proceedings of American Society of Civil Engineers, Vol. 101, No. ST1, January, 1975, pp. 311-324

11. Lennon, T., *Large Compartment Fire Tests*, Proceedings of the 2nd Cardington Conference, 1996

BEHAVIOUR OF GLAZING SYSTEMS IN REAL FIRES

T.J. SHIELDS, S.K.S. HASSANI and G.W.H. SILCOCK
Fire SERT Centre, University of Ulster, UK

1 Introduction

With the advent of fire safety engineering as a recognised professional discipline, the publication of draft Codes of Practice on the application of Fire Safety Engineering Principles and the development of sophisticated quantitative risk assessment techniques it is necessary to be able to select appropriate fire scenarios and input data for fire modelling and risk assessment purposes.

In the past it has been sufficient to assume the instantaneous failure of glazing systems, thereby creating the necessary vent to allow fires to fully develop. In this connection, virtually all the fire growth models in current use assume an $A\sqrt{H}$ factor or equivalent, in some form. However, several factors which interact make it necessary to refine approaches to modelling fire growth:

- thermal insulation regulations reduce convective losses
- developments in glazing techniques influences the performance of the glazing system in real fire environments

Consequently, it is now necessary to be able to accurately predict in real fires, precisely when and where the vent(s) will occur together with the size of vent.

2 Thermal and stress field evaluation

It has been established [1] that the cause of fracturing in glazing when subjected to fire is the non-uniform heating of the glass which produces thermally induced stresses, leading to the initiation of cracking.

Theoretically the breaking stress "$\sigma \propto \Delta T (t)$", where $\Delta T(t)$ is a typical temperature difference between the shaded and unshaded region of the glass at time 't'. The temperature difference ΔT [1] responsible for thermal cracking, is approximated to the temperature rise at the centre of the glass plane, assuming that the edges remain at the initial temperatures.

The temperature rise in the centre of the glass is in turn approximated by either taking the average of both surface temperatures or a thickness averaged temperature rise at the centre of the pane. The temperature difference in the shaded and unshaded region of glass ΔT is related to the induced stresses by the use of simple strain criterion [2, 3].

$$\Delta T = \sigma_b / \varepsilon \beta$$

σ_b = breaking stress
ε = Young's modulus for glass
β = thermal expansion coefficient for glass

Experimental studies [4, 5] have been carried out to validate these models. In [4] a fast fire was produced in a small experimental fire compartment which incorporated a window (280 mm x 500 mm) in the upper section of the compartment wall so that this entire glass pane was immersed in the hot gas layer within ten seconds.

The experimental results in this case were in good agreement with those predicted by the models. However, when large windows (i.e. the glazing extending from upper hot zone to lower cold zone of fire compartment) were used in a series of fifty half-scale fire room experiments[7], the behaviour of glazing differed significantly from that predicted by the models. The lower section of the glass did not fracture for a long time after the initiation of fire. The glass panes remained in-situ for some considerable time and when some part of glass did fall out it occurred only from the upper section of the pane where bifurcated cracks joined to form a network.

3 Stress patterns

It has been established that when window glass is subjected to a thermal field the stress created in the shaded edge is uniaxial tension parallel to the edge, while in the heated area in the vicinity of the edge, stress is uniaxial compression [6]. In the corner region of the glass, stress approximates to an isotropic compression reducing to zero at the physical corner. Therefore, the most likely location for the initiation of cracks is along the edges at some distance from the corners.

Glass edge quality can influence the stress at fracture since the existence of a notch or a chipping at the glass edge increases the stress concentration. It is suggested that the glass edge quality should be categorised in perhaps "Excellent" for as machine cut, ground and bevelled, "Good" for as machine cut only, and "Poor" for as hand cut. For each edge quality classification a factor may then be associated to take account of edge quality on glass breaking strength.

4 Multi-plane glazing units

It is necessary to be able to predict the behaviour of multi-pane glazing in enclosure fires. The extent and the mode of heat transfer from the inner to outer panes before and after the breaking of the first pane is a key factor.

Using mathematical models the performance of double glazed windows in a wild bush fire were predicted [7]. It was suggested that the pane facing the fire would shield the second pane from oncoming heat flux, thereby keeping it cool. If the first pane were to fall out, the second pane would then begin to heat up and subsequently crack. The resulting behaviour of the system is largely expected to depend on whether the first pane would fall out completely or partially. The results of experiments on the performance of double glazed units incorporating ordinary and low emissivity glass carried out at Fire SERT [8] indicated that double glazing units in real enclosure fire, where the conditions have reached flashover in 12 minutes, could retain integrity for over thirty minutes. These findings are in contrast with the common perception that the glazing will probably disintegrate very early in the development of fire. The findings of the Fire SERT experimental programme were in essence confirmed in the last large fires test held here in Cardington on 23rd November 1995. In order to encourage the fire to grow, the double glazing units had to be smashed, in what was already a fairly leaky compartment.

Further, experimental investigation is required to determine how a partially collapsed inner pane, for example, can influence the heat transfer to the outer pane together with the failure mechanisms of the outer pane..

5 Development of experimental method

In the course of preliminary investigation into the behaviour of glazing it became clear that in order to assess the performance of glazing in enclosure fires one must clearly specify the fire characteristics, window geometry and position, both with respect to the room and the fire, and details of the framing. The main parameters that needed to be measured in the assessment of the glazing performance were; heat flux imposed on the glazing, temperature distribution in the glass pane and the associated temperature difference between the shaded and unshaded regions of glass, stress profile in the glass pane, time to first crack, crack bifurcation patterns and extent of glass fall out to assess loss of integrity and insulation.

At Fire SERT, a large fire room which is fitted with a load cell capable of carrying fire loads positioned in the centre or the corner of the room (Figure 1) is used to carry out the glazing experiments. The ventilation of the room is controlled by means of an adjustable vent. The room is instrumented with thermocouples, bi-directional probes, and heat flux meters, to map the temperatures and air flows, etc. One side of the room will accommodate the glazing panels which can take many forms depending on the design of the facade.

Glass surface temperatures on both the fire and ambient sides can be readily measured during the course of fire development by using K type mineral insulated thermocouples with anconal sheaths and junctions near the physical tips. Utilising the flexibility of the sheath material ten thermocouples are formed into coil like arrangements (Figure 2) which makes it possible to position and secure the thermocouple tips on the glass surface, both in the shaded and unshaded region of the glass, and map the temperature distribution in the glazing as fire develops (Figure 3).

By using an appropriate type of strain gauge and taking particular care with their installation (Figure 4) and signal corrections it is possible to measure the dynamic development of thermal strain in the glazing.

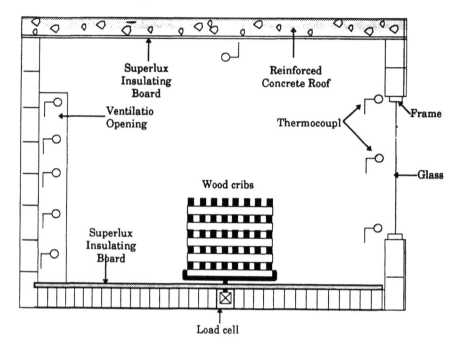

Figure 1.   Large Fire Enclosure With Glazing System and Instrumentation for characterisation of the Fire and the Performance of Glazing.

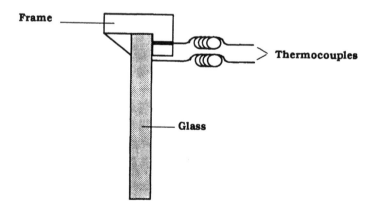

Figure 2.   Details of thermocouples mounted in the shaded and exposed regions of the glass.

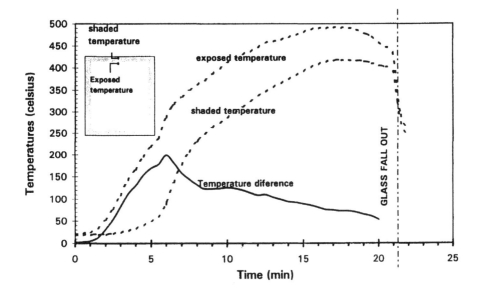

Figure 3.   Local shaded and exposed and the differential temperatures Vs time

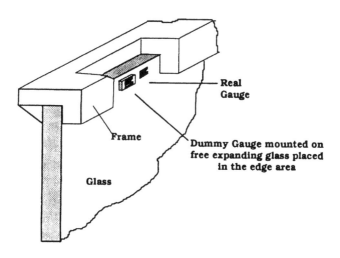

Figure 4.   Details of strain gauges mounted in the shaded region of window glass.

Strain gauge output also accurately indicates the time to the occurrence of cracks in the glass, and the levels of stress at which the cracking occurs (Figure 5).

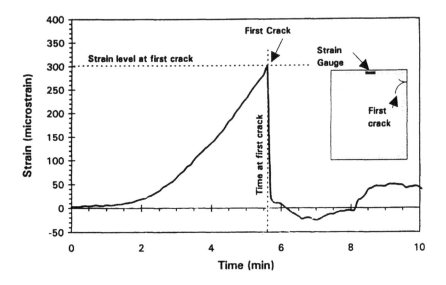

Figure 5.   Strain–time curve indicating strain level and time for first crack

6 Future research

Our current programmes of work is designed to experimentally investigate the mode of heat transfer from fire to glass and to relate this to the development of related thermal stresses.   From stress distributions, it is hoped to be able to predict the crick bifurcation behaviour and ultimately the likelihood and extent of glass failure.   This programme work also includes the application and development of CFD model Sophie to predict the behaviour of glazing in enclosure fires.   The computational side of the project is carried out in collaboration with the Mechanical Engineering Department of Cranfield University.   The aim is to utilise the reliable experimental data to develop a computational model to predict the behaviour of glazing systems in real fire scenarios.

7 References

1.   Emmons, H.W., *The Needed Fire Science*, in Fire Safety Science, Proceedings of the First International Symposium, Grant, C.E. and Pagni, P.J. (Eds), Hemisphere, Washington, D.C., pp 33-35, 1986.
2.   Joshi, A.A., Pagni, P.J., *Users Guide to Break 1, The Berkeley Algorithm for Breaking Window Glass in a Compartment Fire*, Report No. NIST-GCR-91-596, October, 1991.
3.   Pagni, P.J., and Joshi, A.A., *Glass Breaking in Fires*, Proceedings of Third International Symposium on Fire Safety Science, pp 791-802, 1991.

4.   Skelly, M.J., Roby, R.J. and Beyler, C.L. *An Experimental Investigation of Glass Breakage in Compartment Fires*, Journal of Fire Protection Engineering, 3:1, pp 25-34, 1991.

5.   Silcock, G.W.H. and Shields, T.J., *An Experimental Evaluation of Glazing in Compartment Fires*, Proceedings of the Sixth International Interflam Conference, pp. 747-756, 1993.

6.   Blight, G.E., *Thermal Strain and Fracture of Building Glass*, First Australian Conference on Engineering Materials, NSW University, NSW, pp 685-700, 1974.

7.   Cuzzillo, B. and Pagni, P.J., *Windows in Wild Fires*, Wind and Fire II, Abstract of Students Research Project, Couse ME290F, Case Histories, University of California, Berkeley, Spring 1992.

8.   Shields, T.J., Silcock, G.W.H., and Braniff, J.M., *Building Regulation Interaction Report*, 1, Fire Research Centre, University of Ulster, September 1992.

PART TWO

BST–ECSC Fire Programme

A NATURAL FIRE SAFETY CONCEPT FOR BUILDINGS – 1

J.B. SCHLEICH
Principal Engineer, ProfilARBED Recherches, Luxembourg

1 Introduction

The European Commission issued, on 21 December 1988 [9] a directive concerning the products used in the construction of buildings and civil engineering works (Construction Product Directive, CPD). The term 'construction product' refers to products produced for incorporation, in a permanent manner, in the works, and placed as such on the market. It includes materials, elements, and components of prefabricated systems or installations which enable the works to meet the essential requirements. According to that Directive the following essential requirements have to be fulfilled:

- Mechanical resistance and stability
- Safety in case of fire
- Hygiene, health and environment
- Safety in use
- Protection against noise
- Energy economy and heat retention

Concerning 'safety in case of fire', the Directive states that:
"the construction works must be designed and built in such a way that in the event of an outbreak of fire:

- load bearing capacity of construction can be assumed for a specific period of time
- generation and spread of fire and smoke within the works are limited
- spread of fire to neighbouring construction works is limited
- occupants can leave the works or be rescued by other means
- safety of rescue teams is taken into consideration"

For each of these essential requirements, an Interpretative Document was written by a specific Technical Committee of the Standing Committee set up by the EC to follow the implementation of the CPD. In the Interpretative Document 'safety in case of fire'[17], it is foreseen that the essential requirement may be satisfied as far as structural elements are concerned by:

• tests according to harmonised standards or EOTA (European Organisation for Technical Approvals) guidelines, or
• harmonised calculation and design methods, or
• a combination of tests and calculations

Testing methodology standards are mainly developed by CEN TC 127 and calculation methods given in Structural Eurocodes are developed by CEN TC 250. These sets of European standards, which contain the sum of European and world-wide knowledge gathered during the last decades in the field of fire resistance and more specifically on the behaviour of structures in fire, should lead to a uniform manner of assessing the fire resistance of structures throughout Europe. In EC2 to EC6 and EC9, Parts 1.1 deal with design at room temperature and Parts 1.2 deal with structural fire design [18, 20, 26 to 29]. In Part 2.2 of Eurocode 1, the actions in case of fire [25] include both mechanical actions, given by the probable loads applied to a structure during a fire, and thermal actions, represented by the temperature increase in the air and due to a fire.

It is clear that this set of design standards for the fire situation constitutes an important step forward as design methods are allowed to be used instead of fire tests.

2 The need for change

Unfortunately Part 2.2 of Eurocode 2 mainly covers thermal actions arising from the standard temperature-time curve and other nominal temperature-time curves. Physically based (parametric) thermal actions are only dealt with where simplified analytical models or direct design data are available; they are given in informative annexes. It is however known that the conventional or standard fire exposure, according to ISO-834 and section 4.2.2 of ENV1991-2-2, leads to a gas temperature Θ in the fire compartment in °C which depends on the elapsed time 't' in minutes according to the relation

$$\Theta = 20 + 345 \log_{10}(8t+1)$$

This standard temperature-time curve involves an ever increasing air temperature inside the considered compartment, even when later on all consumable materials have been destroyed (figure 1). The application of this standard fire necessarily gives rise to important discrepancies in the fire resistance requirements existing in different countries for the same type of building [15], (figure 2). Furthermore, this so-called 'standard and well defined fire' leads to a warming up of the structural element depending on the type of furnace used in order to perform the ISO fire test. Indeed, radiation conditions inside test furnaces have never been harmonised!

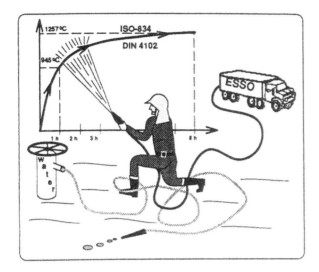

Figure 1.   A possible scenario in order to create ISO-fire conditions

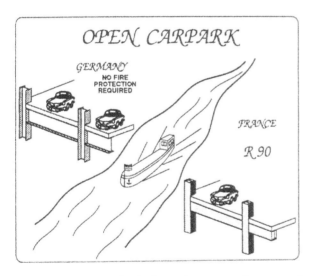

Figure 2.   Discrepancies in fire resistance requirements of neighbouring countries for the same type of building

Consequently it is not astonishing at all, that the very first attempts to define realistic or natural fire conditions have been done even since 1958 [1, 2, 3]. If a natural fire evolution is considered for which active measures have failed, the fully developed fire gives way to a heating phase with increasing air temperatures. However after a given time, depending on the fire load and the ventilation conditions, the air temperatures will necessarily decrease (figure 3).

In order to establish the basis for realistic and credible assumptions to be used in the fire situation for thermal actions, active measures and structural response, a new European Research entitled 'Competitive Steel Buildings through Natural Fire Safety Concept', was started in 1994. It is being performed by 10 partners out of 11 European countries [23] and is coordinated by ProfilARBED Recherches.

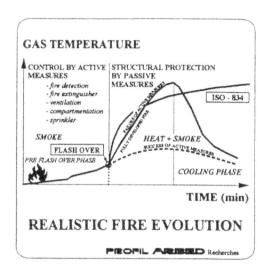

Figure 3.   Natural fire evolution

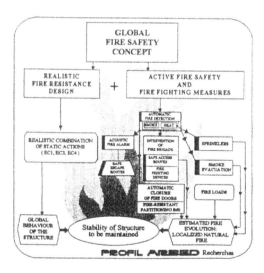

Figure 4.   A Global Fire Safety Concept

The aim is to develop a Global Fire Safety Concept (figure 4) which consists first in a realistic fire resistance design in order to:

- proceed to a global structural analysis in the fire situation
- consider a realistic (accidental) combination rule for actions during fire exposure
- design according to natural fire conditions!

In a second step the global fire safety concept consists in the consideration of the active fire safety and fire fighting measures in view of their impact on the probable evolution of the natural fire.

The first part of this research project is complementary to the demonstration project entitled 'The behaviour of a multi-storey steel-framed building subject to natural fire', and which is coordinated by British Steel Technical [24].

3 Ways and means

3.1 Mechanical actions

As regards mechanical actions, it is commonly agreed that the probability of the combined occurrence of a fire in a building and an extremely high level of mechanical loads is very small. In this respect the load level to be used to check the fire resistance of elements refers to safety factors other than those used for normal design of buildings. The general formula to be used in calculating the relevant effects of actions [25] is:

$$\sum \gamma_{GA} \cdot G_{k,j} + \psi_{1,1} \cdot Q_{k,1} + \sum \psi_{2,i} \cdot Q_{k,i} + \sum A_{d(t)} \tag{1}$$

$G_{k,j}$ = characteristic value of the permanent action ('dead load')
$Q_{k,1}$ = characteristic value of the main variable action
$Q_{k,i}$ = characteristic value of the other variable actions
γ_{GA} = partial safety factor for permanent actions in the accidental situation, 1.0 is suggested
$\psi_{1,1}; \psi_{2,i}$ = combination factors for buildings according to table 9.3, ENV 1991-1 [19]
$A_{d(t)}$ = design value of the accidental action resulting from the fire exposure.

This accidental action is represented by:

- temperature effect on the material properties
- indirect thermal actions created either by deformations and expansions caused by the temperature increase in the structural elements, where as a consequence internal forces and moments may be initiated, either by thermal gradients in cross-sections leading to internal stresses

For instance, in a domestic, residential or an office building with imposed loads as the main variable action ($Q_{k,1}$) and wind or snow as the other variable actions, the formula is

$$1.0\ G_k + 0.5\ Q_{k,1} \qquad \text{since } \psi_2 \text{ for wind and snow are equal to zero} \qquad \text{(F.1.1)}$$

For a storage building the formula becomes

$$1.0\ G_k + 0.9\ Q_{k,1} \qquad\qquad\qquad\qquad\qquad\qquad\qquad\qquad \text{(F.1.2)}$$

When, in a domestic, residential or an office building, the main variable action is considered to be the wind load ($Q_{k,1} = W_{k,1}$) and the imposed load ($Q_{k,2}$ in this case) is the other variable action, the formula is

$$1.0\ G_k + 0.5\ W_{k,1} + 0.3\ Q_{k,2} \qquad\qquad\qquad\qquad\qquad \text{(F.1.3)}$$

In the case of snow as the main variable action ($Q_{k,1} = S_{k,1}$), the formula becomes

$$1.0\ G_k + 0.2\ S_{k,1} + 0.3\ Q_{k,2} \qquad\qquad\qquad\qquad\qquad \text{(F.1.4)}$$

Generally this leads in the fire situation to a loading which corresponds to 50 to 70 % of the ultimate load bearing resistance at room temperature for structural elements.

3.2 Thermal actions

Concerning thermal actions, a distinction is made in Part 2.2 of Eurocode 1 between nominal fires and parametric fires [25].

Nominal fires are conventional fires which can be expressed by a simple formula and which are assumed to be identical whatever the size or design of the building. Nominal fires are mainly (figure 5) the standard fire ISO-834, the hydrocarbon fire reaching a constant temperature of 1100 °C after 30 minutes, and the external fire (used only for external walls) reaching a constant temperature of 680 °C after 30 minutes (4.2 of ENV 1991-2-2). They have to be used in order to prove that an element has the required level of fire resistance to fulfil national or other requirements expressed in terms of fire rating related to one of these nominal fires.

Parametric fires is a general term used to cover fire evolution more in line with real fires expected to occur in buildings. They take into account the main parameters which influence the growth and development of fires.   In this respect the temperature-time curve, and subsequently the heat flux, vary when the size of the building or the amount or kind of fire load varies.   This more realistic way of determining the thermal action due to an expected fire can only be used in association with an assessment by calculation.

Due to the large variety of possible temperature-time curves in a building, the assessment method would have been very expensive if the only possibility was to test components in furnaces for each particular temperature-time fire curve.

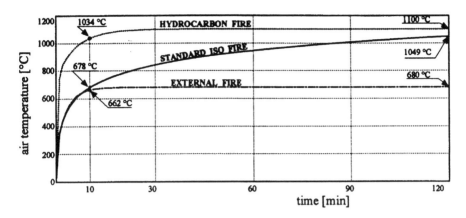

Figure 5. Nominal temperature-time curves according to 4.2 of ENV 1991-2-2 [25].

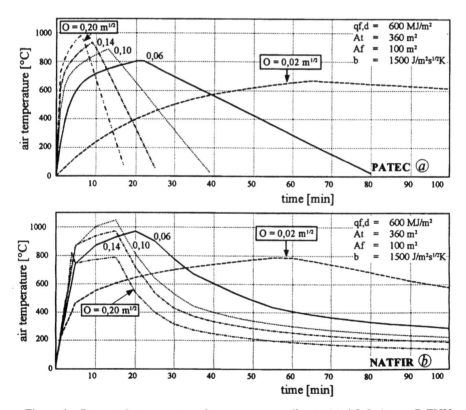

Figure 6. Parametric temperature-time curves according to (a) 4.3 & Annex B ENV 1991-2-2 & (b) following ECSC Research 7210-SA/112, Activity C1 [11]

The present version of Eurocode 1 Part 2.2 gives, for elements inside the building, simplified formulas which take into account the following main parameters: the fire load, the opening factor $O = A_v \cdot \sqrt{h} / A_t$, (with A_v: area of vertical openings, h: height of vertical openings, A_t: total area of enclosure), and the thermal properties of the surrounding walls of the compartment (4.3 and Annex B of ENV 1991-2-2).

An example of the results when using these formulas with a fire load $q_{t,d} = 600$ MJ/m^2, and an opening factor varying from 0.02 m$^{1/2}$ to 0.20 m$^{1/2}$ is shown in figure 6a.   However according to a research founded from 1987 to 1991 by ECSC [11], similar parametric temperature-time curves were established.   Results obtained by this approach, for the same set of previous data, are shown in figure 6b, and seem to be more realistic.   Indeed the heating curves of figure 6a show that the fire is ventilation-controlled [7] for all opening factors from 0.20 m$^{1/2}$ to 0.02 m$^{1/2}$, and that in the cooling phase the temperature-time curve is strictly linear!

On the contrary the heating curves of figure 6b show that the fire is fuel-controlled for opening factors from 0.20 m$^{1/2}$ to 0.10 m$^{1/2}$ and becomes ventilation-controlled for smaller opening factors.   Furthermore, in the cooling phase the temperature-time evolution is curved!

It has to be understood that these nominal or parametric fires correspond to a fully developed fire which means that the corresponding compartment is fully engulfed in fire.   In fact, however, a natural fire may also remain localised which means that the real fire area is limited compared to the total floor area A_f of the compartment.   Normally a two-zone situation is developing in the whole compartment i.e. an upper zone filled with smoke and having higher temperatures, and a lower zone with low temperatures (figure 7).

As long as the lower zone remains sufficiently high, but above all as long as the upper zone gets not a temperature above 500 °C, no flash-over takes place and the natural fire remains localised in the compartment.   Such a situation is given in open car parks, where on one side smoke and heat are escaping easily through the large openings provided in opposite façades, and where on the other side the heat source remains localised in the burning car [15].

This aspect of the evolution of a natural fire is being studied through simple calculation models like two-zone fire models, or more advanced calculation models like multi-zone fire models or computational fluid dynamics (CFD) models [13, 14, 23].   Inside the European Research on the Natural Fire Safety Concept, is Working Group WG1, chaired by Dr J.M. Franssen, which is analysing natural fire models.

3.3 Equivalent time of fire exposure

The following approach, given in Annex E of ENV 1991-2-2 [25], allows to use realistic fire conditions depending on the design fire load density $q_{t,d}$ and on the ventilation, even when the design of members is by tabulated data or simplified rules related to the standard fire (ENV 1992-1-2, ENV 1993-1-2, ENV 1994-1-2).

In fact by definition the equivalent ISO time is the time during which a given structural element has to be submitted to the ISO fire curve in order to obtain, in that element, the same temperature as the maximum temperature produced by the natural fire curve.   It was when applying this principle to concrete cross-sections, with reinforcing bars protected by a 30 mm thick concrete layer, that equation (E.1) was established.

$$t_{e,d} = q_{f,d} \cdot k_b \cdot w_f \qquad\qquad (E.1)$$

The equivalent ISO time $t_{e,d}$, formulated in this way with k_b and w_f given in (4) and (5) of Annex E, is material independent, but $t_{e,d}$ should in fact be material dependent.

Indeed when using Annex B, giving parametric temperature-time curves according to equations (B.1) to (B.6), and applying these natural heating curves as well as the standard fire to different cross-sections, the non linear finite element code CEFICOSS allowed to establish the differential and transient temperature fields in those cross-sections [35].

The conclusion drawn when considering the previous definition of the equivalent time was:

- The two methods, Annex E and Annex B together with CEFICOSS, lead to similar equivalent times for concrete cross-sections or sections made of protected steel profiles.
- However if the cross-section is an unprotected steel profile, these two methods give contradictory results. In fact equation (E.1) of Annex E gives too high values for $t_{e,d}$, or finally leads to a much too severe heating up of the steel section under a given natural fire. Indeed,

> if A_v (or O) $\nearrow$. $t_{e,d}$ $\nearrow$ according to Annex B
> but $t_{e,d}$ $\searrow$ according to Annex E

Therefore equation (E.1) may be improved to apply to unprotected steel:

$$t_{e,d} = \left(q_{f,d} \cdot k_b \cdot w_f \right) k_c \qquad\qquad (E.1.1)$$

with k_c correction factor function of the material composing structural cross-sections and defined in Table 1.

Table 1.   Correction factor K_c, to be applied to the equivalent time $t_{e,d}$ of Annex E [25] in order to cover various cross-sections. (O is the opening factor defined in Annex B in $m^{1/2}$)

Cross section material	Correction factor (k_c)
Reinforced concrete	1.0
Protected steel	1.0
Non protected steel	$13.7 \cdot O$

For composite construction elements, the equivalent time as defined before, based on a unique temperature equivalence, is not valid anymore. Indeed such a procedure would depend on the considered element, beam or column, and on the considered point in the cross-section. When applying this method to the composite frame tested under natural fire conditions in Braunschweig (April 12 and 24 1989) [10], it was found that the equivalent ISO time $t_{e,d}$ would scatter from 42 to 80 minutes (figure 8).

In fact for composite structures the equivalence between a natural fire and the ISO-fire has to be based on the equivalence of the load bearing capacity. This was done for the previously named composite frame test [10], and the equivalent ISO time $t_{e,d}$ obtained was 46 minutes (figure 9). This means that this is the time after which that composite frame, submitted to the ISO-fire, has the same bearing capacity as the minimum bearing capacity produced by the natural fire. Inside the European Research on Natural Fire Safety Concept is Working Group WG2, chaired by Dr U Kirchner, which is analysing the equivalent time of fire exposure.

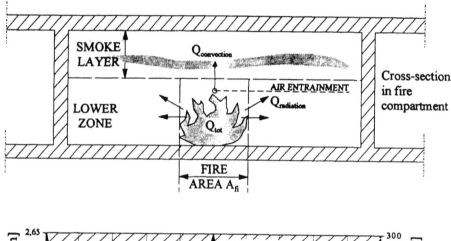

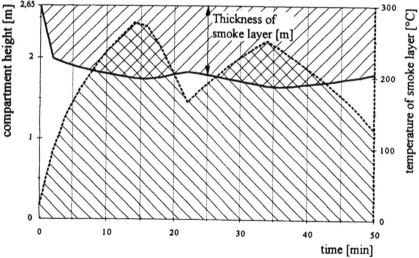

Figure 7.   Calculation by the two-zone and multi-compartment fire model HAZARD [22] of the thickness of the smoke layer and its temperature in function of time; the fire source is given by a burning car

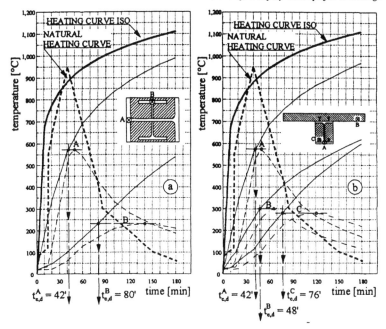

Figure 8. Temperatures calculated by CEFICOSS in the composite column a and
the composite beam b for the ISO heating (——) or the natural heating
(- - -) according to the Braunschweig tests on April12 and 24, 1989 [10].

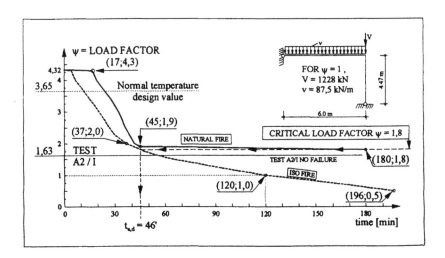

Figure 9. Composite frame A2-I / Fire resistance calculated by CEFICOSS in
function of ψ for the ISO-fire and natural fire defined in Figure 8 [10].

3.4 Fire Characteristics

In order to define physically-based thermal actions, the input data required for a natural fire design have to be identified. These are:

- the fire load density
- the combustion model, the entrainment of air into the fire plume, the rate of heat release, the thermal radiation, the fire size and fire spread and eventually the flash-over
- the building geometry, the opening factor and thermal inertia of surfaces surrounding the compartment

In the framework of this presentation only, aspects dealing with the fire load density and the rate of heat release will be discussed.

3.4.1 Characteristic Fire Load density

The fire load Q in a fire compartment is defined by the total energy liable to be released. Building components such as wall and ceiling linings, and building contents, such as furniture constitute the fire load. Divided by a reference area (generally the floor area) the fire load Q gives the fire load density q_f. The fire load density is the source of the fire development and is also the production-source term in the heat balance equation solved by calculation models.

In the Part 2.2 of Eurocode 1, the characteristic fire load density is defined by the following equation:

$$q_{k,\ell} = \frac{1}{A_f} \sum_i \psi_i m_i H_{ui} M_{ki} \qquad \text{in } [MJ/m^2]$$

M_{ki} = the characteristic mass of the combustible material i [kg]
H_{ui} = the net calorific value of the material i [MJ/kg]
m_i = the factor describing the combustion behaviour of the material i
ψ_i = the factor assessing that the material i is a protected fire load
A_f = the floor area of the fire compartment
$H_{ui}M_i$ represents the total amount of energy contained in material i and released assuming a complete combustion.

The m-factor is a non-dimensional factor between 0 and 1, representing the combustion efficiency: m = 1 corresponds to complete combustion and m = 0 to the extreme situation in which a combustible material procures no heat at all to the fire process. The m-factor is a function of the type of fuel (solids, liquids) and its geometrical properties (porosity, massivity), its position in the fire compartment (exposed area to radiation) and the fire characteristics (temperature, oxygen content, etc.). So far, no international agreement exists on the way to determine the m-factor. Of course, the m-factor may be assumed conservatively as m = 1.

The ψ-factor is introduced to take into account protection of the fire load, for example by putting it inside a cabinet. It has a value between 0 (complete protection for the full fire duration) and 1 (the protection has no influence on the energy release). The protection may reduce the energy release, but often does so for a limited period of time, which depends on the fire conditions (radiation, temperature). This is not reflected in the present concept of the ψ-factor which is time independent. For many practical applications $\psi = 1$ is a realistic value.

According to Part 2.2 of Eurocode 1 the characteristic fire load density in a compartment may be obtained by performing a fire load survey based on the above equation. The net calorific value H_{ui} of the corresponding combustible material may be taken from existing literature [4, 21, 25].

This Eurocode 1 option for a specific study of the fire load may be extended to give a general procedure as follows:

- The net calorific value of numerous materials is known. If necessary, H_{ui} can be determined on the basis of a generally accepted test method (calorific bomb, ISO 1716).

- In addition, the m-factor should be known. As mentioned earlier however, an internationally accepted test method is not available. A concept of such a method is under development within the scope of the European Research on the Natural Fire Safety Concept [23]. Basic ideas behind that work are, that the conditions under which the m-factor is determined should follow real fire conditions closely, and that modern measuring techniques should be applied. Until such a method is available, a conventional value for the m-factor of 0.7 is proposed.

The characteristic fire load density $q_{k,f}$ may also be determined from a national fire load classification system, based on the occupancy of the corresponding fire compartment (Table 2). This type of information is available in Sweden [3], Switzerland [5] and United Kingdom [21], and may also be found in the CIB/W14 workshop report [4] or in the ECCS Design Manual [6].

Table 2. Fire load density $q_{k,f}$ for different occupancies [21].
     * 80% fractile is the value not exceeded in 80% of the room occupancies.
     + Storage of combustible materials.

Fire load densities (MJ/m^2)		
Occupancy	Average	Fractile * 80%
Dwelling	780	870
Hospital	230	350
Hospital storage	2000	3000
Hotel bedroom	310	400
Offices	420	570
Shops	600	900
Manufacturing	300	470
Manufacturing/Storage + <150 kg/m²	1180	1800
Libraries	1500	2250
Schools	285	360

3.4.2 Rate of Heat Release

An essential parameter in a fire is the rate of heat release (RHR). It is the source of the gas temperature rise, and the driving force behind the spreading of gas and smoke. A typical fire starts small and goes through a growth phase. Two things may happen:

1. During the growth process there is always enough oxygen to sustain combustion, in which case, when the fire size reaches a maximum, the RHR is limited by the available fire load (fuel controlled fire).
2. The size of openings in the compartment enclosure is too small to allow enough air to enter the compartment, in which case the RHR is limited by the available oxygen and the fire is said to be ventilation controlled.

Both ventilation and fuel controlled fires can go through flashover. This important phenomenon marks the transition from a localised fire to a fire involving all the exposed combustible surfaces in the compartment.

The fire model of Annex C in Part 2.2 of Eurocode 1 determines whether the fire is fuel or ventilation controlled, and then uses the appropriate relationship to calculate the RHR.

$$R = \min\left(\frac{L}{\tau_F}; \quad 0{,}18\left(1 - e^{-0.036\eta}\right)A_v\sqrt{h.W/D}\right) \qquad (C.6)$$

R	$=$	rate of burning [kg of wood/s]; N.B. 1 kg wood $\cong$ 17 to 20 MJ
L	$=$	total fire load [kg of wood]
τ_F	$=$	free burning fire duration (assumed to be 1200 s)
A_v	$=$	sum of window area on all walls $\left(A_v = \sum_i A_{v,i}\right)$; [m²]
h	$=$	weighted average of window height on walls $\left(h = \sum_i A_i h_i / A_v\right)$; [m]
W	$=$	width of wall containing window(s) [m]
D	$=$	depth of fire compartment [m]
η	$=$	$A_T / A_v \sqrt{h}$; [m$^{-1/2}$]
A_T	$=$	all surfaces minus windows $(A_t - A_v)$; [m²]

The first term in equation (C.6) relates to fuel controlled fires and the second to ventilation controlled fires. The expression for fuel control is only approximate, as it assumes that all fuel controlled fires have a duration of twenty minutes.

For the ventilation-controlled regime, various sets of equations exist for the RHR but the differences are small. This can be seen in graphical form in figure 10, which compares expressions for RHR from different sources with experimental data [21, 25].

Pre-flashover fires are not dealt with in the Eurocode, but they are important in egress analysis and in the design of heat and smoke extraction systems. They will also gain relevance for structural fire safety when the concept of localised fire is accepted. For pre-flashover fires one method of calculating the RHR uses conventional or calculated values of the RHR per unit area, and estimates the rate of fire spread (i.e. the rate of growth of the area involved in fire) to calculate the RHR as a function of time.

Values of the RHR depending on the building occupancy [21] or on the heat and smoke venting systems [16] vary from 250 to 500 kW/m².

Inside the European Research on the Natural Fire Safety Concept is Working Group WG3, chaired by Dr L Twilt, which is analysing the aspects related to fire characteristics in general [34].

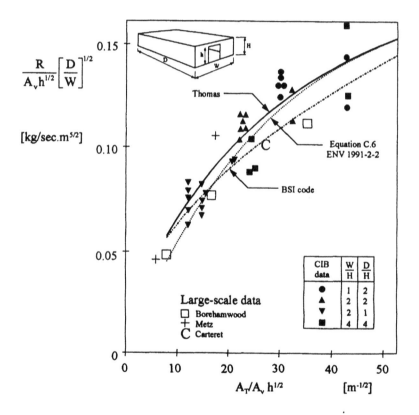

Figure 10. Variation of the heat release R in [kg of wood/s] with compartment size data (D; W; A_T) and ventilation data (h; Av) for ventilation controlled fires [21, 25].

3.5 Sprinkler models and statistics

Water, widely available and inexpensive, is the most commonly used fire extinguishing fluid. It should, however, not be used in case of hydrocarbon fires, as hydrocarbons will float on top of the water, continue to burn and even spread following the water layer.

According to RP Fleming [31], automatic sprinkler systems are considered to be the most effective and economical way to apply water to suppress a fire. The wet pipe system is by far the most common type of sprinkler system. It consists of a network of piping containing water under pressure. Automatic sprinklers are connected to the piping such that each sprinkler protects an assigned building area. The application of heat to any sprinkler will cause that single sprinkler to operate, permitting water to discharge over its area of protection.

Researchers at the National Institute of Standards and Technology (NIST) developed, in 1993, a model of the effectiveness of sprinklers in reducing the heat release rate of furnishing fires based on measurements of wood crib fire suppression [12]. The model assumes that all fuels have the same degree of resistance to suppression as a wood crib, despite the fact that tests have shown furnishings with large burning surface areas can be extinguished easily compared to the deep-seated fires encountered with wood cribs.

The recommended equation gives the variation of the rate of heat release R_t [kW] for any time t following the activation of sprinklers t_{act} [s], and in function of the spray density $\dot{w}$ [mm/s] which gives the uniform water layer discharged in one second by the water distribution nozzles

$$R_t = R_{tact} \cdot e^{\left[(t_{act} - t)(\dot{w})^{1.85}/3\right]}$$

The reduction of the rate of heat release is illustrated in figure 11 for a maximum rate of heat release R_{act} of 500 kW, existing when sprinklers get activated. Moreover the effect of the spray density is demonstrated for $\dot{w}$ varying from 0.08 mm/s to 0.2 mm/s. It can be seen that for a spray density of 0.08 mm/s or 48 mm/min, which corresponds to $4.8l/m^2 \cdot min$, the rate of heat release and consequently the fire size gets down to 196 kW after 5 minutes of sprinkler operation. The fire may be considered as suppressed after more or less 20 minutes.

A quite interesting research has just been completed by the 'Studiengesellschaft' in Germany [32] showing that sprinklers are able to suppress a fire even when high air temperatures have been obtained. In this case a 100 kg wood cible established on a platform of $1 m^2$ was put to fire in a furnace of $23 m^2$ horizontal cross-section and 2.55 height. The sprinklers were activated 8 minutes after ignition.

According to figure 12 the decreasing temperatures clearly indicate that the fire has been suppressed after 22 minutes of sprinkler activation producing an application density of $4.4l/m^2 \cdot min$.

Automatic sprinklers however serve a dual function as both water distribution nozzles and heat detectors. This last function, similar to smoke detectors, allows the automatic fire detection and subsequent automatic fire alarm.

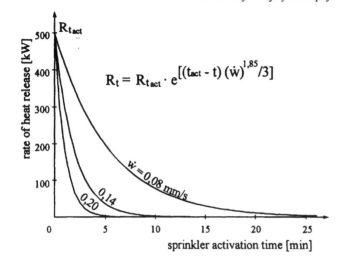

Figure 11. Suppression of fire i.e. reduction of rate of heat release [kW] by sprinkler sprays of various application densities w [mm/s], [31].

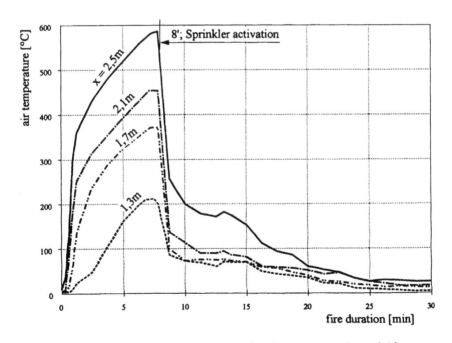

Figure 12. Suppression of fire i.e. reduction of the air temperature by sprinkler spray density of 4.4l/m² · min at various heights in the fire compartment [32].

The state of the art on sprinkler action requires now to thoroughly analyse existing data on real fires. Inside the European Research on the Natural Fire Safety Concept is Working Group WG4, chaired by Professor M Fontana, which is studying sprinkler models and statistics.

4 Towards a global fire safety concept

The Global Fire Safety Concept described earlier, 'The Need for Change' consists of the following four main aspects:

1. fire safety of occupants and firemen
2. required natural fire resistance period
3. realistic structural fire design including active fire safety
4. reduction of material losses.

4.1 Fire safety of occupants and firemen

Usually in the case of a fire the fatalities or injuries are mainly due to the smoke and toxic gases.

The first goal of a Fire Safety Concept is to protect occupants against smoke inhalation by providing an early detection, enabling people to escape from the critical area, also activating smoke venting to reduce the smoke layer or activating a sprinkler system to suppress the smoke production and to protect the escape routes; this means safety first.

In fact smoke evacuation shall be a must for all new and refurbished buildings as well for any existing building of a given size. This is a "Conditio sine qua non" for the delivery of construction permissions.

It should be checked that the potential risk is limited as follows:

Potential Risk ≤ Acceptable Risk X Protection Factor

Certain Risk Assessment Methods [5,8] give a quantification of this protection factor, which expresses a safety improvement as shown in Table 3. If smoke detectors are installed and the fire detection automatically activates the smoke evacuation, the safety of occupants is increased by a factor of 1.7.

Table 3.   Protection factors increasing the safety of occupants in face of smoke [5, 8] and the safety of firemen in face of heat.

Protection factors	Smoke detection installed	Automatic smoke evacuation	Global protection from smoke	Safety of
ANPI (B)	$U = 1.05^8 = 1.48$	$U = 1.05^3 = 1.16$	1.71	occupants
SIA (CH)	$s_{,,} = 1.45$	$s_e = 1.20$	1.74	
Proposal	Compartmentation R60 $A_t \leq 1000 \, m^2$ $1.05^4 = 1.21$	Protected access staircase R60 $1.05^4 = 1.21$	Global protection from heat 1.47	Safety of firemen

Table 4.   Safety factor γ_{s2} covering the consequences of a structural failure when calculating the required natural fire resistance period.

Compartment floor area A_f (m^2)	Proposed safety factor γ_{s2}			
	One storey building n = 1	Two storey building n = 2	Building	Building
≤ 2500	1.00	1.25	1.50	2.00
2500 < A_f ≤ 5000	1.05	1.40	1.75	2.50
5000 < A_f ≤ 10000	1.10	1.50	–	–
10000 < A_f ≤ 20000	1.20	1.60	–	–

Similar safety considerations should be made for firemen. Table 4 shows a proposal for their safety improvement in case of a convenient building compartmentation and for a protected staircase permitting a safe access route to fight the fire.

4.2 Required natural fire resistance period

The required fire resistance period under natural fire conditions, $t_{fi,requ}^{nat}$, should be dependent on the objectives to avoid any human fatalities and to reduce the consequences of a structural failure. This could be expressed by:

$$t_{fi,requ}^{nat} = x' \cdot \gamma_{s1} \cdot \gamma_{s2} \qquad (1)$$

x' =   equal to a minimum fire resistance of 15 minutes,

γ_{s1} =   safety factor proportional to the evacuation time needed for the specific occupants,

γ_{s2} =   safety factor related to the consequences a structural failure would have on the neighbouring environment, on the building content etc.

The safety factor γ_{s1}, proportional to the evacuation time of occupants, may be a function of:

(p)   the correction coefficient for the time of evacuation, function of the mobility of occupants,

    1. Independent and mobile (workers) $p = p' = 1$

    2. Mobile but dependant (school kids) $p = p' = 2$

    3. Immobilised people (illness) $p = p' = 8$

    4. No compartmentation between the floors $p = p' + 2$

    5. No plan of evacuation $p = p' + 2$

    6. Risk of panic $p = p' + 2$

(E)   the floor level number of the compartment (ground floor = 0, first floor = 1, first basement floor = -1). The most critical compartment will generally be chosen from the top level or the deepest basement level.

(X)   the number of people to be evacuated from the compartment,

(L)   is the length of the compartment (the longest distance between the centres of opposite walls of a compartment), [m]

(b)   is the width of the compartment (A_r = L.b), [m]

(x)   is the number of exit routes from the compartment. The minimum width of an exit is 600 mm but account should be taken of the local circumstances i.e. in a hospital, exit should be wide enough for the free passage of hospital beds.

The proposed relation is given as

$$\gamma_{s1} = p \; [(E/10)^2 + 1] \; [3,2 \cdot 10^{-8} \; (X \; (L+b) \, / \, x^2)^2 + 8] \qquad (2)$$

Values of γ_{s1} obtained by this equation are shown in figure 13 for different types of buildings and various sets of the previously explained parameters, and may vary from γ_{s1} = 0,82 for industrial halls, up to γ_{s1} = 10,4 for hospitals.

The safety factor γ_{s2}, related to the consequences a structural failure would have on the neighbouring environment or on the building content may be a function of the floor area A_r of the compartment and may depend on the number of storeys.

Values of γ_{s2} are proposed in Table 4 and may vary from γ_{s2} = 1,0 for a one-storey building up to γ_{s2} = 2,5 for high-rise buildings.

Figure 13 shows the effect of γ_{s1} and γ_{s2} on the evaluation of the required natural fire resistance period which may vary from 13 to 300 minutes.

Under certain circumstances the required fire resistance $t_{fi,requ}^{nat}$ may be replaced by the requirement that the building shall never collapse! Indeed the consequences of a structural failure may be considered as unacceptable for economical reasons (computer centre, head office of a company,...), for social reasons (high-rise tower in a city centre,...), or for historical reasons (museum,...). Values of $t_{fi,requ}^{nat}$ obtained for hospitals and high-rise buildings are large enough to be equivalent to the requirement "No failure under natural fire conditions".

Evaluation	VARIOUS BUILDING SITUATIONS / n = number of storeys					
of γ_{s1}	INDUSTRIAL HALL n = 1	SCHOOL n = 4	OFFICE BUILDING n = 11	HOSPITAL n = 8	HIGH RISE BUILDING n = 31	DANCING n = 2
p	1	4	1	8	1	3
E	0	3	10	7	30	1
X	20	300	50	60	100	1000
L	100	60	50	70	50	60
b	50	20	30	30	50	30
x	2	4	2	2	2	4
γ_{s1}	0,82	3,8	1,7	10,4	10,0	5,5
γ_{s2}	1,05	1,5	2,0	1,5	2,0	1,25
$t_{fi,requ}^{nat}$	13'	86'	51'	234'	300'	103'

Figure 13. Required natural fire resistance period for various building situations.

4.3 Realistic structural fire design including active fire safety

It is essential to consider the expected natural fire conditions, function of the active fire safety and fire fighting measures. This may be done, according to equation (D.1) of Part 2.2 of Eurocode 1 [25], through the design fire load density $q_{f,d}$ defined as:

$$q_{f,d} = \gamma_n \cdot \gamma_q \cdot q_{k,f} \quad \text{(D.1)}$$

γ_n = differentiation factor ≤ 1, accounting for active measures

γ_q = safety factor normally ≥ 1, depending on the consequences of failure and frequency of fires

$q_{k,f}$ = characteristic fire load density [M J/m^2] as explained in chapter 3.4.1

The differentiation factor γ_n, function of the various fire safety and fire fighting measures able to reduce the practical heating effect of the characteristic fire load density, may be given by

$$\gamma_n = \gamma_{n1} \cdot \gamma_{n2} \cdot \text{.......} \cdot \gamma_{n8} \cdot \gamma_{n9} \leq 1 \quad \text{(D.1.1)}$$

Referring to various national regulatory documents [5, 8, 30] and following ENV1991-2-2, figure 14 was established showing that γ_n could be split up in three main contributions due to the fire extinction effect of approved sprinklers (γ_{n1} and γ_{n2}), to the automatic fire detection and alarm transmission (γ_{n3} to γ_{n5}), and to the intervention of the fire brigade (γ_{n6} to γ_{n9}).

It follows from this figure that γ_n may vary between 0.07 and 0.54

The safety factor γ_q, depending on the consequences of failure and the frequency of fires, may be given by

$$\gamma_q = \gamma_{q1} \cdot \gamma_{q2} \text{ normally} \geq 1 \, \text{(D.1.2)}$$

Table 5.   Safety factor γ_{q2} according to SIA 81 [5] in function of the building occupancy.

Safety factor γ_{q2}	Danger of fire activation	Examples of occupancies
0.85	small	art gallery, museum
1.00	normal	residence, hotel, paper industry
1.20	mean	manufactory for machines/engines
1.45	high	chemical laboratory, paint workshop
1.80	very high	manufactory of fireworks or paints

The safety factor γ_{q1}, related to the consequence of structural failure, is function of the size A_f of the compartment under fire and of the number of storeys of the building. It could be considered identical to γ_{f2} given in Table 4.

The safety factor γ_{q2}, related to the frequency of fires, is depending on the danger of fire activation and therefore is function of occupancies as proposed in Table 5.

γ_{ni} Function of Active Fire Safety Measures

Official Document		Automatic Water Extinguishing System γ_{n1}	Independent Water Supplies 0 | 1 | 2 γ_{n2}	Automatic fire Detection & Alarm by Heat γ_{n3}	by Smoke γ_{n4}	Automatic Alarm Transmission to Fire Brigade γ_{n5}	Work Fire Brigade γ_{n6}	Off Site Fire Brigade γ_{n7}	Safe Access Routes γ_{n8}	Fire Fighting Devices γ_{n9}	$\gamma_n^{min} = \gamma_{n1} \cdot \gamma_{n2} \cdots \gamma_{n9}$ $\gamma_n^{max} = \gamma_{n1} \cdot \gamma_{n9}$
Title	Date of publication										
SIA 81 (CH)	1984	0,50 / 0,59	/	0,83 or 0,69		0,83	0,67 or 0,63 / $\gamma_{n6} \cdot \gamma_{n7}$ = 0,53		/	1,0 / 1,39*	0,15 / 0,49
ANPI (B)	1988	0,58 ⊙	1,0 or 0,86 or 0,65	0,82	0,68	included in ⊙	0,50	0,68	/	1,0 / 1,36*	0,07 / 0,48
DIN 18230-1 Project	1995	0,60	/	0,90		/	0,60	/	/	/	0,32 / 0,54
ENV 1991-2-2	1995	0,60	/	/	/	/	0,60	0,70	/	/	0,60
PROPOSAL NFSC	1996	0,60	1,0 or 0,9 or 0,7	0,90	0,80	0,80	0,60	0,70	1,0 / 1,5*	1,0 / 1,5*	0,10 / 0,54

Figure 14. Differentiation factor γ_n accounting for various active fire safety measures γ_{ni} as given in different regulatory documents [5, 8, 25, 30] and as proposed in the European Research on the Natural Fire Safety Concept [23, 36]; values of γ_{n8} and γ_{n9} larger than 1.0 apply if the active measure is insufficient.(*)

By considering γ_n and γ_q factors as described here, the design fire load density $q_{f,d}$ given by equation (D.1) may be reduced down to 15 % of the characteristic fire load density. This effective fire load density shall be taken in order to estimate the realistic thermal action in one or several fire compartments, considering of course also realistic thermal boundary conditions.

This fire exposure together with the accidental combination rule for actions (equation F.1), shall be considered when performing a global structural analysis[23, 36]. The design fire resistance shall exceed the required natural fire resistance period.

$$t_{fi,d}^{nat} \geq t_{fi,requ}^{nat} \qquad (3)$$

4.4 Reduction of material losses

This Global Fire Safety Concept is also the best way to reduce, in a considerable manner, material losses, as well direct losses as losses due to business interruption [33].

Indeed the implementation of a consistent set of fire safety and fire fighting measures reduces the development of a fire, minimises the potential of flashover, and improves the response from the fire brigade. As losses are presently quite high, according to the Association of British Insurers (figure 15), the use of this G.F.S.C. may consequently be supported by insurers, as insurance payouts will drop.

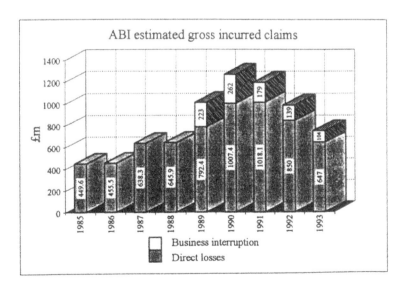

Figure 15. Losses recorded by the Association of British Insurers

5 Conclusion

The previously described Global Fire Safety Concept [23] is in total compliance with the Construction Product Directive [9] concerning 'safety in case of fire'. Indeed all fire sub-requirements given in the Introduction to this paper may be fulfilled:

1. *"the load bearing capacity of the construction can be assumed for a specific period of time";* this is covered by fulfilling equation (3)

$$t_{fi,d}^{nat} \geq t_{fi,requ}^{nat}$$

2. which in turn requires on one side the activation of 'Mechanical Actions' and thermal actions, and on the other side the consideration of active fire safety measures. An estimation of the required natural fire resistance period $t_{fi,requ}^{nat}$ is proposed in section 'Towards a Global Fire Safety Concept'.

3. *"the generation and spread of fire and smoke within the works are limited";* this is considered in section 'Realistic Structural Fire Design' through the differentiation factor γ_n accounting for active measures as proposed in figure 14.

4. *"the spread of fire to neighbouring construction works is limited";* this is considered in section 'Realistic Structural Fire Design' through the safety factor γ_q depending on the consequence of failure and frequency of fires (Table 4 and 5).

5. *"occupants can leave the works or be rescued by other means"* and *"the safety of rescue teams is taken into consideration";* this is covered in section 'Fire Safety of Occupants and Firemen' when defining protection factors in Table 3, and in section 'Required Natural Fire Resistance Period' through the safety factor γ_{s1} related to the evacuation time needed for the specific occupants.

Furthermore it has to be underlined that this *Global Fire Safety Concept* is structured around the following three priorities:

- consider safety first for occupants and firemen,
- define the required natural fire resistance period $t_{fi,requ}^{nat}$ such as to avoid any human fatality and to reduce the consequences of a structural failure,
- design a fire resistance of the structure exceeding the required natural fire resistance period.

6 References

1.   Kawagoe K.; *Fire behaviour in rooms.* Japanese Building Research Institute, Report N° 27, Tokyo, 1958.

2.	Kawagoe K., Sekine T.; *Estimation of fire temperature-time curve in rooms.* Japanese Building Research Institute, Occasional Reports N° 11 and 17, Tokyo, 1963/64.

3.	Petterson O., Magnusson S.E., Thor J.; *Fire Engineering Design of Steel Structures.* Publ. 50, Swedish Institute of Steel Construction, Stockholm, 1976.

4.	Thomas P.H. & Co., *Design Guide Structural Fire Safety.* Workshop CIB W14, January 1983.

5.	SIA; Brandrisikobewertung, Berechnungsverfahren. Dokumentation SIA 81, Zürich, 1984.

6.	ECCS-TC3; Design Manual on the *European Recommendations for the Fire Safety of Steel Structures.* Publ. N° 35/ECCS, Brussels, 1985.

7.	Wickström, U.; *Application of the standard fire curve for expressing natural fires for design purposes.* ASTM STP 882, pp. 145-159, 1985.

8.	ANPI; *Evaluation des risques.* Association Nationale pour la Protection contre l'Incendie, Ottignies, 1988.

9.	EC; Construction Product Directive, dated 21.12.1988, 89/106/EEC. Official Journal of the European Communities, N° L40/12, Luxembourg, 11.02.1989.

10.	Schleich J.B., Lickes J.P.; *Simulation of test frames A2/I and A2/II.* ECSC Research 7210-SA/112, Activity A2/B1 / RPS Report N°04/90, Luxembourg, 26.11.1990.

11.	Schleich J.B., Scherer M.; *Compartment temperature curves in function of opening factor and fire load.* ECSC Research 7210-SA/112, Activity C1 / RPS Report N°08/90, Luxembourg, 02.02.1991.

12.	Evans D.D.; *Sprinkler Fire Suppression Algorithm for HAZARD,* NISTIR 5254. National Institute of Standards and Technology, Gaithersburg, MD, 1993.

13.	Schleich J.B., Cajot L.G.; *Natural fires in closed car parks.* ECSC Research 7210-SA/518 etc., B-E-F-L-NL, 1993-96.

14.	Schleich J.B., Cajot L.G.; *Natural fires in large compartments.* ECSC Research 7210-SA/517 etc., B-E-F-L-NL, 1993-96.

15.	ECCS-TC3; *Fire Safety in Open Car Parks.* Technical Note N° 75/ECCS, Brussels, 1993.

16.	NBN; NBN S21-208 / "*Protection incendie dans les bâtiments* - Conception et calcul des installations d'évacuation de fumées et de chaleur - Partie 1: Grands espaces intérieurs non cloisonnés s'étendant sur un niveau". Institut Belge de Normalisation, Brussels, August 1993.

17.	EC; Interpretative Document N°2, Essential Requirement *Safety in case of fire.* Official Journal of the European Communities, N° C62/23-72, Luxembourg 28.02.1994.

18.	CEN; ENV 1994-1-2, Eurocode 4 - *Design of composite steel and concrete structures,* Part 1.2 - Structural fire design. CEN Central Secretariat, Brussels, DAV 30.10.1994.

19.	CEN; ENV 1991-1, Eurocode 1 - *Basis of design and actions on structures,* Part 1 - Basis of design. CEN Central Secretariat, Brussels, DAV 30.10.1994.

20.	CEN; ENV 1995-1-2, Eurocode 5 - *Design of timber structures,* Part 1.2 - Structural fire design. CEN Central Secretariat, Brussels, DAV 30.11.1994.

21.   BSI; *The Application of Fire Safety Engineering Principles to Fire Safety in Buildings*. Draft British Standard Code of Practice, London, June 1994.
22.   Peacock R.D. & Co.; *An update guide for HAZARD* I, Version 1.2, NISTIR 5410. National Institute of Standards and Technology, Gaithersburg, May 1994.
23.   Schleich J.B.; *Competitive steel buildings through natural fire safety concept*. ECSC Research 7210-SA/522 etc., B-D-E-F-I-L-NL-UK & ECCS, 1994-98.
24.   Martin D. M.; *The behaviour of a multi-storey steel framed building subject to natural fires*. ECSC Pilot Project 7215-CA/806 etc., F-NL-UK, 1994-97.
25.   CEN; ENV 1991-2-2, Eurocode 1 - *Basis of design and actions on structures*, Part 2.2 - Actions on structures exposed to fire. CEN Central Secretariat, Brussels, DAV 09.02.1995.
26.   CEN; ENV 1992-1-2, Eurocode 2 - *Design of concrete structures*, Part 1.2 - Structural fire design. CEN Central Secretariat, Brussels, Final Draft Sept. 1995.
27.   CEN; ENV 1993-1-2, Eurocode 3 - *Design of steel structures*, Part 1.2 - Structural fire design. CEN Central Secretariat, Brussels, Final Draft July 1995.
28.   CEN; ENV 1996-1-2, Eurocode 6 - *Design of masonry structures*, Part 1.2 - Structural fire design. CEN Central Secretariat, Brussels, Final Draft June 1995.
29.   CEN; ENV 1999-1-2, Eurocode 9 - *Design of aluminium alloy structures*, Part 1.2 - Structural fire design. CEN Central Secretariat, Brussels, First Draft May 1995.
30.   DIN; Baulicher Brandschutz im Industriebau - Teil 1: Rechnerisch erforderliche Feuerwider-standsdauer. DIN 18230-1, Beuth Verlag GmbH, Berlin, 1995.
31.   Fleming R.P.; *Automatic Sprinkler System Calculation*. SFPE Handbook of Fire Protection Engineering, Section 4 / Chapter 3, NFPA Publication, Quincy-Massachusetts, June 1995.
32.   Klaus J., Kohl K.J., Kutz M.; Rechnerisch/experimentelle Untersuchungen zur Erfassung des Einflusses von Massnahmen, zur Verhinderung der Brandausbreitung u. zur Brand-bekämpfung, auf die Brandraumtemperaturen-twicklung bei Naturbränden. Studiengesellschaft, Projekt 224, Düsseldorf, 1995.
33.   Scoones K.; FPA *Large Fire Analysis* 1993. Fire Prevention Issue 286, Borehamwood, January / February 1996.
34.   Twilt L. & Co; *Input data for the natural fire design of building structures*. IABSE Colloquium, Basis of Design and Actions on Structures, Delft, 27/29.03.1996.
35.   Franssen J.M. & Co.; *Parametric temperature-time curves and equivalent time of fire exposure*. IABSE Colloquium, Basis of Design and Actions on Structures, Delft, 27/29.03.1996.
36.   Cajot L.G., & Co.; *Influence of the Active Fire Protection Measures*. IABSE Colloquium, Basis of Design and Actions on Structures, Delft, 27/29.03.1996.

A NATURAL FIRE SAFETY CONCEPT FOR BUILDINGS – 2

D.M. MARTIN
Swinden Technology Centre, British Steel Plc,UK

1 Introduction

During the last decade a considerable number of fire resistance tests have been carried out in many European countries. Most of these tests have, however, been limited to evaluating performances of individual structural members in compartments with carefully controlled gas fire atmospheres. Tests of this type are useful for establishing the behaviour of steel members in a standard test but they cannot represent natural fire scenarios since they do not allow for the restraining effects of other structural members present in the building, nor do they adequately represent the thermal environment of a natural fire. This is particularly true in assessing the fire resistance of modern multi-storey steel-framed buildings where general load redistribution, floor membrane action and connection behaviour significantly improve structural fire performance over that achieved by testing single elements.

Indeed, there is already significant evidence that the fire resistance of whole structures is much better than that of single elements on which fire resistance is usually assessed. This evidence has arisen from various sources including 3-D numerical model predictions, real fires such as that of Broadgate [1] and also large scale tests conducted in Australia [2, 3].

In view of this, British Steel is leading a large pan-European research and demonstration project which concerns the behaviour of the Cardington LBTF eight-storey steel-framed test structure when subjected to natural fires. Other collaborators include:

TNO Building & Construction Research (NL); Centre Technique Industriel de la Construction Metallique (CTICM) (F); The Steel Construction Institute (UK) and the University of Sheffield (UK). Funding is being provided by British Steel plc, the European Coal and Steel Community, TNO and CTICM.

The overall objective of this project is to develop a greater understanding of the natural fire resistance of multi-storey steel-framed buildings. It is clear for example, that not all of the members of a steel-framed structure should require the same degree of fire protection. The degree of restraint, likelihood of loss of strength or stiffness under fire attack and the consequences of failure will vary significantly depending upon the position and function of the particular element within the building structure. In this project we seek to quantify these effects and so bring about the basis for a new, rational, natural fire safety concept for steel-framed buildings. This will lead to significant increases in safety, economy and competitiveness for this modern form of construction.

2 Objectives of the project

The specific objectives of the project may be summarised as follows:

1. To demonstrate the true behaviour of natural fires in a real multi-storey steel-framed building, and to assess the effects of those fires upon the structural steel frame and floors.
2. To validate and develop existing analysis procedures for the behaviour of structural steel frameworks under elevated temperature conditions.
3. To provide guidance based upon validated computer models on the effective design of steel structures to prevent collapse in fire.
4. To contribute towards the development of an integrated natural fire safety philosophy for steel-framed buildings.

An important concept which underpins much of the above is acceptance that multi-storey building structures possess a significant degree of structural redundancy and so they are able to safely redistribute loads when individual elements suffer local losses in stiffness or strength.

3 Project management

The technical planning and direction of the project is overseen by a Steering Committee which is chaired by the author.   British Steel Technical is entirely responsible for carrying out the large scale fire tests and in this context Dr B.R. Kirby acts as the project leader and supervises the work of four task groups:
    Mechanics and instrumentation; compartment design and construction; software and data acquisition and fire engineering design.
    Dr M.A. O'Connor is responsible for the numerical modelling work undertaken within British Steel Technical and co-ordinates this activity within the project as a whole, working closely with colleagues at TNO, CTICM and also with Professor Plank at the University of Sheffield.
    The Steel Construction Institute are engaged under contract to British Steel Technical and will prepare design guidance based upon the project findings.   The project is also being closely co-ordinated with two other major areas of investigation:

- *Natural fire safety concept.* This is a large risk and hazard assessment study and the representatives from every country of the European Community are involved. The project is being led by Arbed Recherche, Luxembourg. The UK are involved as a main partner.
- *Behaviour of the BRE frame subject to fire attack.* This work is being carried out by the BRE and involves two large scale fire tests on the multi-storey steel framed building structure. These tests have been planned to complement those being carried out under the British Steel programme and the technical staff from the various organisations work together in planning and preparing for each test.

The Project Managers for the three research programmes (led by British Steel, Arbed Recherches and the Building Research Establishment) work closely together to ensure a complementary approach. It should be noted that the total value of these projects exceeds £3 million and all three should be substantially complete by December 1997.

4 Experimental work

The programme of work involves four major fire tests within the steel-framed multi-storey building structure. Three of these large scale tests have now been completed and the extensive data (from thermocouples, strain gauges, displacement transducers, inclinometers, etc.) is currently being analysed and documented. The general behaviour of the steel may be summarised as follows:

4.1 Test 1
The first test (on the steelwork supporting the seventh floor) involved heating a 9 m spanning composite beam by means of a gas-fired furnace. The end beam-column connections consisted of simple partial depth end plates. No fire protection was applied to the beam. The steel section and associated composite slab were heated to temperatures approaching 870 °C over a period of $2^3/4$ hours. The maximum deflection of the beam at the end of the test was only some 230 mm, i.e. span/35, indicating excellent inherent fire resistance.

4.2 Test 2
The second test was the largest gas-controlled furnace test ever undertaken in a real building. The steelwork supporting the fourth floor of the building was heated across a 2.5 m strip across the full 21 m width of the building. Two internal and two edge columns together with three composite beams (6 m, 9 m and 6 m spans) were included in the test. All six beam-column joints were simple partial depth end plates whereas the beam-beam connections were fin plates. The columns were passively fire protected to within 200 mm of the lower flanges of the beams.

Maximum steel temperatures of the order of 820 °C were achieved over a period of 2 hours. In this case the maximum floor deflection experienced was approximately span/34 again indicating good inherent fire resistance in the absence of passive protection. In the later stages of this test (at steel temperatures of around 750°C) the two internal columns suffered severe local distortions in the exposed connection areas; similar to that observed at Broadgate [1]. The external columns suffered no significant deformations or damage. The feasibility of utilising unprotected columns and/or connections will depend upon the particular restraint conditions (e.g. position within the structure) together with the applied load level and joint design aspects.

4.3 Test 3

The third test involved a 10 m x 7.5 m corner compartment of the structure, and, unlike the previous gas-fired tests, heat was applied (to the steelwork supporting the second floor) by means of a natural wooden crib fire. The mass of these wooden cribs amounted to some 45 kg/m^2 which, in terms of calorific value, is considered to be at the higher end of the typical office spectrum. As with the previous tests static structural loading of around 50% of the design live loading was applied by means of sandbags. Steel temperatures in excess of 1000 °C were achieved over a period of 1 hour 25 minutes. The maximum floor deflection measured during the test (some recovery took place during the cooling down period) was 365 mm, i.e. span/25. It should be noted that the increases in floor deflection with temperature occurred in a very stable manner with no abrupt deteriorations in stiffness. Effectively, an equivalent time of fire exposure (as compared with the ISO time-temperature curve) of 1$^{1}/_{2}$ hours was achieved.

4.4 Test 4

The final fire test will involve approximately $^{1}/_{4}$ of the entire plan area of the second floor and real office furniture (carpets, desks, chairs, filing cabinets etc.) will be used as the fire loading. Guests will be invited from all countries of the European Community to witness this test which will take place in June 1996.

5 Numerical work and design guidance

The results of these major tests are being utilised to correlate finite element models (ABAQUS, DIANA, SISMEF, LENAS, INSTAF) which are currently under development. The correlated numerical models will be utilised to broaden the database of useful results and enable a great variety of structural configurations and fire attack scenarios to be considered.

6 Conclusion

1. There is now an acceptance that the fire resistance (in terms of structural response) of whole structures under natural fire loading is significantly better than that of single elements on which fire resistance is universally assessed.
2. Three large co-ordinated research projects are currently underway in Europe with the aim of bringing about a more rational approach to assessing the fire resistance of multi-storey steel framed buildings. These projects include structural - thermal loading and response as well as risk and hazard assessment.

3. In the British Steel programme three major fire tests in the BRE multi-storey steel framed building have now been completed.   The test results indicated that composite floor beams possess a very significant degree of inherent fire resistance which may lead to the elimination of passive fire protection requirements for such members.

4. The situation in the case of column elements appeared to be less favourable and severe local buckling in certain exposed connection areas occurred in the second fire test.   This aspect is the subject of further work particularly as regards connection design, restraint conditions and the level of the applied column load at the fire limit state.

5. The final British Steel test will take place in June 1996.   It will be the largest of the tests undertaken to date and will involve real furniture as the fire loading.

6. Good progress is being made on the numerical analysis work using the ABAQUS, LENAS, DIANA, SISMEF and INSTAF codes.   These models will be fully correlated prior to carrying out parametric studies and preparing design guidance.

7 References

1.   *Investigation of Broadgate Phase 8 Fire*, The Steel Construction Institute, Ascot, UK.

2.   *Fire Test of the 140 William Street Office Building*, Broken Hill Proprietary Company Ltd., Melbourne, Australia.   Report BHPR/ENG/12/92/043/SG2C.

3.   *The Effect of Fire at 140 William Street*, Broken Hill Proprietary Company Ltd., Melbourne, Australia.   Report BHPR/ENG/12/92/044/SG2C.

BRITISH STEEL TECHNICAL EUROPEAN FIRE TEST PROGRAMME – DESIGN, CONSTRUCTION AND RESULTS

B.R. KIRBY
Swinden Technology Centre, British Steel Plc, UK

1 Introduction

At the Cardington LBTF during 1995, three fire tests on the eight-storey steel-framed building were carried out by British Steel Technical as part of a European funded programme. These broke new grounds in the technical and logistical planning and implementation of fire research studies.

This paper describes in detail the design and construction of the test arrangements, the instrumentation installed, as well as some of the initial results and their implications upon future designs. Fire engineering aspects concerning the third test are discussed including initial plans for the fourth test in the series, a demonstration office fire.

As project leader to this programme, one of my roles has been to try to bring together the thoughts and ideas of the structural engineers and modellers into designing tests that examined the more important issues and providing suitable data for developing and validating models.

2 Test programme

In the programme being conducted by British Steel Technical, four tests were planned. Generally, these are referred to as:

- I-D Restrained Beam Test   Figure 1
- Plane Frame Test   Figure 2
- Corner Test   Figure 3
- Demonstration (Furniture) Test   Figure 4

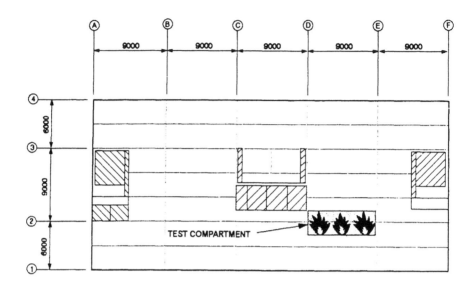

Figure 1.   Restrained beam test

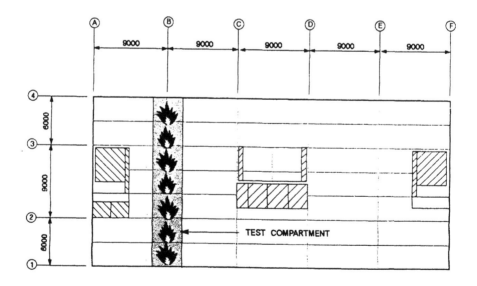

Figure 2.   Plane frame test

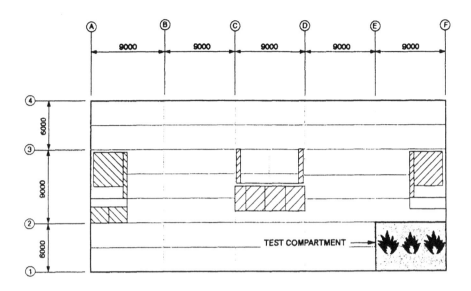

Figure 3.   Corner compartment test

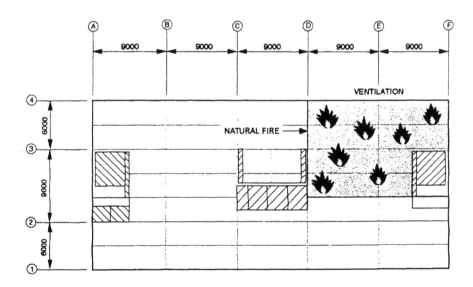

Figure 4.   Demonstration test

Each test was designed to evaluate different parts of the frame in which either a single beam or series of beams and columns were heated artificially, or, using wood or furniture to represent a real fire condition. You will also see from the size and geometry of the test compartments the complexity of each test increases as the programme progresses, although technically and logistically the most demanding to date has been the 2-D Plane Frame Test.

2.1 Restrained beam

The first test required heating a single 305 x 165 mm beam and the surrounding concrete floor spanning 9 m between a pair of 254 x 254 mm column sections (Figure 5). The steelwork around the connections also had to remain cool.

For the test, the mode of heating was not important provided it could be controlled and an accurate history of the temperature distribution in the structure was obtained. Considerable time was therefore spent in deliberating on the type of heating fuel which should be used, the choice being either gas, electricity or wooden cribs (real fire).

Following checks on the gas supplies available at Cardington it was decided to build a gas-fired furnace using in-house expertise from our own fuel and furnaces department. Since time and space on the eight-storey frame is at a premium and the test beam was located at Level 7, a modular system was designed and fabricated off-site. This entailed building a wire cage 8 m long x 3 m wide x 2 m high insulated with mineral wool and ceramic (Figure 6).

Prior to constructing the furnace, the test beam and surrounding structure was extensively instrumented with strain gauges, thermocouples, position sensors and inclinometers for measuring rotation. Areas of the concrete slab were also instrumented to measure temperature distributions through the floor and strain within the reinforcing mesh. In total, nearly 300 separate pieces of instrumentation were installed.

Although it is not current practice for meeting fire resistance/protection requirements, the voids between the top flange and metal decking were filled with mineral wool. This procedure was however carried out to reduce the temperature gradients and make modelling the first test easier. In subsequent tests the voids remained unfilled.

Using sand bags, each weighing 11 kN, the floor and surrounding structure was loaded to a level corresponding to full dead + $^1/_3$ live which is typical of the service load found in normal office buildings. This level of live loading is semi-permanently maintained throughout the building.

In the test, the beam was heated at a rate of 3–10 °C per minute until temperatures of 800–900 °C were achieved (Figure 7). At this point the mid-span deflection had reached 230 mm ($^L/_{35}$)and the test terminated. Figure 8 is a plot of temperature Vs mid-span deflection which shows that the beam, even at nearly 900 °C, had still not attained runaway when the test was terminated. In comparison with a BS 476 fire resistance test, the limit of stability would have been attained at around 700 °C. This therefore reflects the influence of the continuity and slab behaviour on the overall stability of the exposed beam when forming part of a composite frame.

During heating, the web and lower flange at each end of the beam buckled just inside the furnace (Figure 9).

Figure 5.   General view of the structure to be tested

Figure 6.   Test furnace in position

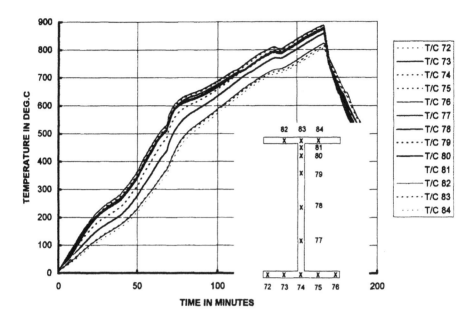

Figure 7.   Test 1: Temperature profile through beam during heating phase

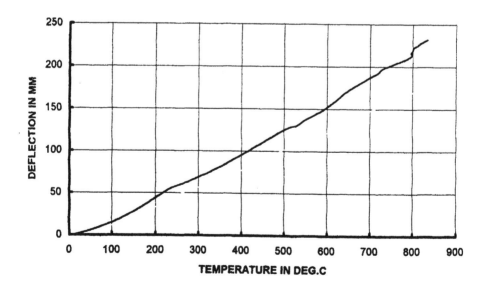

Figure 8.   Test 1: Steel temperature Vs mid-span deflection

Figure 9.   Test 1: Local buckling of test beam just inside furnace

Figure 10. Test 1: Local deformation in lower flange at end of test beam

There was also evidence of the expansion of the lower flange impinging upon the end plate (Figure 10). These type of effects can be identified by changes in strain gauge readings (Figure 11).

One effect that we did want to look at was the result of expansion of heated members on movement in the structural frame. Whilst accepting the test was conducted on a single internal beam, the resulting displacement between the two columns was less than 3 mm (Figure 12).

2.2 2-D Plane frame test

The second test conducted in May '95 involved heating a series of beams and columns across the full 21 m width of the building, taking in both the primary and secondary members (Figure 13).

Having been successful with using gas in the first test it was decided to use gas for the second also. However, it was recognised that there were major technical and logistical problems, for instance:

1. The furnace would be 21 m long and therefore nearly 5 times longer than typical floor furnaces used for conducting BS 476 (ISO 834) fire resistance tests on floor beams. It was perhaps the largest furnace ever constructed on a building.
2. The design of the entire furnace system would be based upon theoretical considerations.
3. It had to work first time.
4. The furnace construction, test and demolition, together with installation and dismantling the instrumentation, had to be completed within four weeks.

A furnace 21 m long x 3 m wide x 4 m high was constructed on the third level using 190 mm lightweight concrete blockwork. To improve the efficiency of the heating system this was lined with 50 mm ceramic fibre blanket. For heating, eight industrial burners were installed along one side and these were served by a 3" gas main and an 8" air supply (Figure 14).

One of the objectives of the test was to examine the behaviour of the frame around the connections, and the type of deformation observed in the Broadgate fire. For this reason, all the columns were protected with a thin insulation layer up to 200 mm below the lower flanges of the beams. The beams as well as the beam/column and beam/beam connections however, remained totally unprotected. A total of 600 pieces of instrumentation were installed.

In order to replicate what occurred in the Broadgate fire, the test was to a point where considerable deformation occurred in and around the connections, and well beyond that normally reached in the Standard Furnace Test (Figures 15 and 16).

At the end of the test, the primary beams had attained a temperature of 820 °C in the lower flange with a maximum deflection of span/34. Temperatures in the exposed portions of the columns had achieved 750 °C.

2.3 3-D Corner test

The third test in the programme was carried out in July and used some of the lessons learned from the previous two tests.

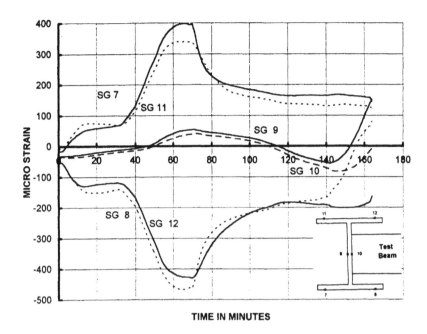

Figure 11. Test 1: Strain measurements in column below test beam connection

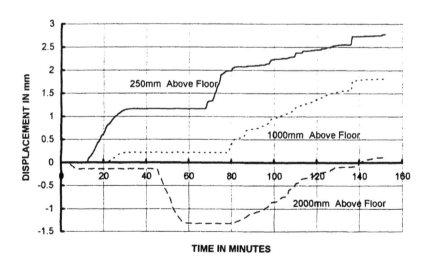

Figure 12. Test 1: Horizontal displacements between columns above test floor

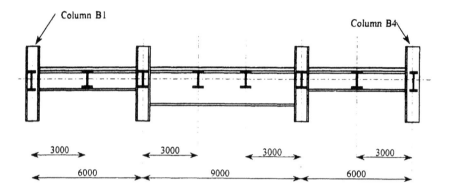

Figure 13. Test 2: Schematic representation of structural frame to be tested

Figure 14. Test 2: Part of the 21m furnace

Figure 15. Test 2: General view of the frame after the test

Figure 16. Test 2: Gross deformation around beam/column connection

Using concrete blockwork, a compartment measuring approximately 10 m x 7.5 m was built within the corner of the frame, with the main objective of determining the extent of membrane action provided by the composite metal deck floor system (Figure 17). The columns were protected over their entire length and around the connections using ceramic fibre. Protection was also provided to the edge beams, however, all the internal primary and secondary beams including the beam to beam connections were left exposed to fire attack.

A fire loading of 3400 kg of wood was used to provide a fire load density of 45 kg/m^2 of floor. This amount of combustible material exceeds that found in the majority of offices and the 38 kg of wood/m^2 density representing the 80% fracture proposed in design codes for office buildings.

In order to achieve a relatively slow atmosphere heating rate, and heat the unprotected steel members to temperatures of between 900 °C and 1000 °C, ventilation through the original opening in the facade was considerably reduced (Figure 18). In addition, while theoretical calculations were made of the likely fire process, it was deemed prudent to provide some control over combustion rates by installing a screen. This could be raised or lowered during the test.

Overall, the temperatures achieved by the unprotected beams were generally between 900 °C and 1020 °C, with a maximum deflection of 420 mm over a 9 m span. Using indicative specimens to characterise the natural fire, the severity was equivalent to a period of heating in the BS 476 fire resistance test, of just under 90 minutes.

Figure 17. Test 3: Internal view of test compartment nearing completion

Figure 18. Test 3: Front view of compartment during early fire development stage

Although the main emphasis of this work has been concentrated on the structural response, valuable information has been gained which will enable us to extend the study on large compartment fires carried out in 1993 with the Fire Research Station [1].

2.4 Demonstration – furniture test

The fourth test in the programme, due to take place in June 1996, will be a fire in a simulated open plan office.

Precise details at the time of preparing this paper have not been finalised since they will, to a large extent, depend upon analyses of the previous results. The compartment will, however, occupy a floor area of around 140 m^2 and incorporate modern furniture fittings and computers etc.

3 Materials testing

Running in parallel with the fire tests, a programme of materials testing is also being carried out at both ambient and elevated temperatures. Apart from a check on the properties of the steel members and components used in the construction of the frame, it provides a database for detailed structural analysis at the fire limit state.

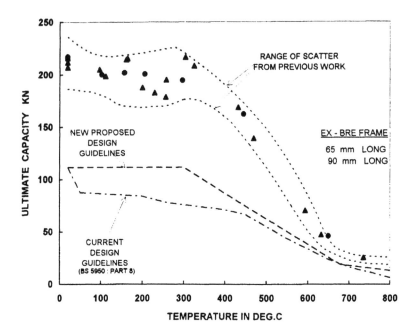

Figure 19. Elevated temperature test results on fully threaded bolts used in the frame, compared with existing data

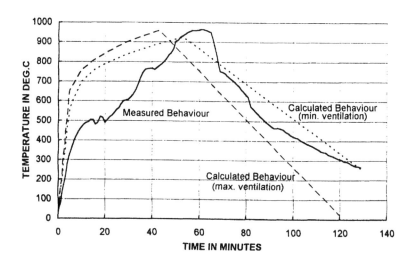

Figure 20. Comparison between measured and predicted atmosphere time/temperature response

For example, Figure 19 illustrates the results of a series of elevated temperature tests carried out on fully threaded bolts used at the connections, and compares them with data from a recent, more extensive, study [2].

4 Conclusion

British Steel Technical has considerable expertise in carrying out and analysing the results of major fire test programmes and have often worked closely with other organisations in developing the subject of fire engineering.

These tests, which were the first to be carried out on such a large purpose built frame, have broken new grounds in technical and logistical planning and the implementation of fire research studies:

- In each test, the construction of the furnace/compartment, installing the instrumentation, fuel supplies, carrying out the test followed by dismantling the entire set up, all had to be completed within four weeks in which up to $1^{1}/_{2}$ man years of effort was expended.
- The second test which involved the construction of a furnace 21 m long x 4 m high, was almost five times larger than a BS 476 floor beam furnace. It also had to work first time.
- In each test up to 600 separate pieces of instrumentation were installed requiring 30 km of cabling to connect back to the data loggers.
- From the point of view of the results, while it will be some time before the analysis will be completed, some important aspects are emerging which will have an impact on future designs.
- Steel temperatures approaching 900 °C in Test 1 and 1000 °C in Test 3 have demonstrated the continuity provided by the surrounding structure and membrane action in the floor slab when one or more members lose stiffness and strength.
- Test 2 demonstrated that in real structures very extensive deformations can be tolerated by steel members without structural collapse.
- Thermal expansion of individual members can be constrained by the surrounding structure resulting in negligible movement of the building's external facade.
- In test 3, comparing the measured time temperature response of the atmosphere with the predicted behaviour using the parametric equation in Eurocode 1 provided a reasonable conservative correlation (Figure 20). This is particularly important for the further development and use of the Eurocodes during the ENV period.

5 References

1.   Kirby, B. R., Wainman, D. E., Tomlinson L. N., Kay, T. R. and Peacock, B. N., *Natural Fires in Large Scale Compartments* - A British Steel Technical Fire Research Station Collaborative Project. British Steel Technical, Swinden Technology Centre, 1994.

2.   Kirby, B. R., *The Behaviour of High Strength Grade 8.8 Bolts in Fire.* J.
        Construct. Steel Research, 33 (1995) 3-38.

NUMERICAL MODELLING OF COMPOSITE STRUCTURES SUBJECT TO THERMAL LOADING

M.A. O'CONNOR
Swinden Technology Centre, British Steel Plc, UK

1 Introduction

Experimental testing of structures subject to thermal loading is an extremely expensive business. In other branches of structural analysis and design, cost benefits have been demonstrated in adopting the dual approach of numerical modelling with selective full-scale testing. The full scale tests are used to examine the fundamental structural behaviour so that numerical models can be developed and calibrated. Once the numerical model is fully calibrated, it can be used to generate further data thus extending the database of available results.

Numerical modelling and experimental testing of steel structures subject to thermal loading has tended to concentrate on the behaviour of individual structural elements. However, since the Broadgate fire of 1990, it has been recognised that there are inherent benefits in considering the behaviour of structural frameworks in fire as a whole rather than as a collection of individual elements. Therefore, attention is now switching to the behaviour of these individual structural elements in frameworks.

The move to modelling framework behaviour in fire, particularly composite behaviour, is not straightforward as there are different problems to be tackled to those found in modelling individual element behaviour. This paper outlines some of the problems that need to be overcome to model successfully the behaviour of composite steel frameworks subject to thermal loading.

2 Composite framework behaviour subject to thermal loading

Composite steel framework behaviour under thermal loading can be addressed on two levels. These are the local behaviour of the individual structural elements in the framework, and the global behaviour of the overall framework. Obviously the effect of the local structural element behaviour affects the global framework behaviour, but the distinction is important as the time to failure of an individual element is strongly influenced by the whole framework.

The local structural element behaviour in fire is governed by three main factors. These are the temperature distribution in the structural element, the load carried by the structural element and the degree of restraint offered by the surrounding structure. A rise in temperature of a structural element will lead both to expansion of the element and degradation in the material properties of that element. If the structural element is significantly restrained by the rest of the structure then the expansion will lead to an increase in the stresses in the element. This increase in stress along with the degradation of material properties can lead to failure of the local element. If an individual structural element fails then the load carried by that element will need to be redistributed to the surrounding structure. The extent of the redistribution will depend on the temperature of the surrounding structural elements and the degree of redundancy in those elements.

3 Numerical modelling requirements

The most important aspect to be modelled by any developed numerical model, therefore, is the amount of redistribution that can take place in the composite structure before possible structural collapse of any part of the structure. Structural collapse is defined as the inability of the structure or part of the structure to carry the design loading in a controlled manner. This means that any developed numerical model needs to be able to simulate the correct load path through the structure and include the various modes of failure of the individual structural elements. This behaviour has to be simulated throughout the thermal loading phase and, possibly, the unloading phase.

As well as the technical requirements of any numerical model, there are certain practical limitations to be placed on the development of the model. These practical limitations include the ease of use of the model (both in pre and post processing) and the hardware requirements of the model. For ease of use the developed model ideally needs to be set up using a graphical user interface. To enable hardware requirements to be kept to a minimum, the number of nodes and elements, the element types used to represent the behaviour and the solution procedures used need to be as computationally efficient as possible. The number of nodes and elements used can be kept to a minimum by only modelling the thermally affected part of the structure. The element types used will be governed by the modes of failure that need to be modelled for each structural element type. The solution procedures needed will be governed by the failure mode of the individual elements and the overall structure.

The main elements to be modelled are columns, beams and floor slabs. The connections between these main structural elements will also be important to some degree. It may, however, be difficult and the matter of some judgement to separate connection behaviour from the main structural element behaviour.

Failure of the main structural elements can take place in a variety of modes. This depends upon the loading in the element, the type of the element and the restraint conditions applied to the element. It will be important to set up any model so that the correct conditions are obtained in each structural element throughout the analysis. This is not as straightforward as might first appear due to the complex interaction that can take place when structural elements are acting compositely. The behaviour that may need to be included to model composite framework behaviour for each structural element is outlined below.

3.1 Column behaviour

The behaviour of columns in fire is relatively simple to model compared to the behaviour of the composite beam and floor slab. Columns are only restrained at the end of each column. The main possible column failure modes are squash failure, overall buckling failure, bending failure or lateral torsional buckling failure. It is relatively straight forward to follow the stress pattern due to static and thermal loading throughout the element using beam-column theory. Therefore, by following the spread of yield across the cross-section, it is possible to predict all of these failure modes using one-dimensional beam elements.

However, it may also be possible to get a local buckling failure of a column in extreme circumstances. This failure mode cannot be modelled using existing beam-column theory and would entail modelling the cross-sectional behaviour of the column using shell elements for the flanges and web of the column. This is computationally very expensive if a large extent of the model is thermally loaded.

3.2 Beam and slab behaviour

The behaviour of the beam and floor slab has to be considered together due to the composite interaction between the elements and the dual function of the floor slab. The dual function of the floor slab is to enhance the load carrying capacity of the bare steel beam and to act as the main structural element that carries the floor loading back to the surrounding structure. This raises fundamental questions as to the modelling approach that needs to be adopted in modelling both beam and slab behaviour.

There can be two approaches to modelling beam and floor slab behaviour. The first approach is to separate the dual functions of the slab. That is, to model the behaviour of the floor slab acting compositely with the beam separately from the floor slab's load carrying function. The second approach is to model the dual function of the slab directly by assuming anisotropic properties for the slab that would give the correct overall behaviour when coupled with the bare steel element. The first approach whilst possibly being simpler than the second is not ideal owing to the apriori assumptions that have to be made about the interaction between the floor slab and beam during thermal loading.

The main failure modes of the composite beam that will need to be predicted are squash failure and bending failure. Overall buckling modes are not so important as these are restrained by the composite action of the floor slab. These failure modes can be modelled using simple beam-column theory. However, distortional lateral buckling failure may be important for more slender beam elements, particularly in hogging moment regions of the composite beam. This failure mode can be quite localised and may require to be modelled using shell elements for the flanges and webs of the steel beam.

The modelling of the floor slab should concentrate initially on load-carrying capacity but ultimately the prediction of overall failure will be important. Initially, floor slab behaviour will be governed by the flexural stiffness in the main load carrying direction. As the steel beams supporting the floor slab increase in temperature and lose stiffness, the floor slab has to span effectively over a longer distance. This leads to loss of flexural stiffness and the membrane stiffness of the slab becomes more important.

The need to model both the flexural and membrane stiffnesses of the floor slab means that the behaviour needs to be modelled using shell elements. This leads to difficulties in numerical modelling due to the lack of knowledge of modelling non-linear anisotropic shells. These difficulties are compounded by the limited concrete models available for reinforced concrete in tension, making floor slab behaviour computationally very expensive to model.

3.3 Connection behaviour

The main connections that affect composite frame behaviour under thermal loading are beam to column, beam to beam and beam to floor slab connections (shear studs).

Of these connections, the degree of shear interaction throughout the thermal loading is perhaps the most difficult to quantify. This will have to be dealt with by considering upper and lower bound solutions. That is by considering that there is full and no shear interaction between the beam and the floor slab throughout the thermal loading.

As regards connections between the beam and other beams and columns, it is considered that the beam to column connection will have the greater influence on behaviour. This, of course, will be strongly dependent upon the type of connection adopted. Consideration has to be given to the axial restraint supplied to the end of the beam as well as the (more normal) degree of rotational restraint. This behaviour can be modelled by using separate nodes for the beam and column connected by spring elements.

4 Numerical modelling study

The possible requirements of a numerical model to study composite structures under thermal loading have been outlined above. Some of these requirements are computationally extremely onerous. Therefore, as part of the ECSC project, a numerical study is being undertaken to ascertain the influence of the various factors on overall framework behaviour. This study is being undertaken by a numerical modelling group consisting of British Steel Technical (UK), TNO (NL), CTICM (F) and Sheffield University (UK). The first three partners are developing numerical models using general purpose finite element programs so that full allowance can be made for all possible influences on the framework behaviour. Sheffield University are using a 3-D framework program that is being developed in-house.

The numerical models are being developed using the results of the first major fire test, i.e. the restrained beam test. This was the simplest test carried out but it included all the major influences that need to be modelled to simulate composite frame behaviour under thermal loading. The major influences to be studied are:

- The extent of the frame to be modelled
- The effect of local buckling on composite beam behaviour
- Concrete floor slab modelling
- Beam to column connection behaviour
- The degree of composite shear interaction
- The influence of temperature profiles through the main structural elements

5 Conclusion

1. The aim of the numerical modelling group is to develop a numerical model that can adequately address all the likely failure modes of composite steel frames under thermal loading.
2. As discussed above, the behaviour of composite steel frameworks in fire is complicated with the local behaviour of individual structural elements significantly affecting the behaviour of the overall structure.
3. If this behaviour is to be successfully modelled on a practical basis, then some simplifying assumptions need to be made about the way in which the various structural elements interact during fire. Some of likely requirements of numerical models to model the behaviour have been outlined above.
4. As part of the ECSC project, a rational study on the relative importance of these requirements on overall structural behaviour is being carried out.

ANALYTICAL ASPECTS OF THE CARDINGTON FIRE TEST PROGRAMME

C. BOTH, R.J. VAN FOEKEN, and L. TWILT
TNO Building and Construction Research, The Netherlands

1 Introduction

During the last decade, numerical models have been used more and more to analyse the behaviour of structures under fire action. In view of the limited dimensions of the test furnaces, the validation of these numerical models is mainly based on fire tests on single structural components. Furthermore, it is common practice, at least in Europe, to exclude structural restraint in fire tests. The reason is that structural restraint is hard to accomplish and differs from case to case.

In real structures, there will often be, to some extent, structural restraint, and, as a consequence, the single-element approach is conservative.

The ECSC co-sponsored project "The behaviour of a multi-storey steel-framed building subject to fire", offers unique possibilities to validate numerical models on the basis of full-scale fire tests on complete structural systems, including restraint effects. These fire tests are carried out at the Large Building Test Facility at Cardington.

This paper deals with the application of the general purpose finite element package DIANA.

The potential of this FEM package to describe the thermal and mechanical response of fire exposed structural elements and systems is illustrated on the basis of numerical analyses of well defined fire tests on composite steel/concrete structural elements, and a more complex, though academic, structural system.

Main attention is on the evaluation of the first two Cardington fire tests: the restrained beam test and the plane frame test.

2 Single element validation

2.1 Thermal response of composite steel/concrete slab

In the scope of the ECSC co-sponsored research project "Fire resistance of composite concrete slabs with profiled steel sheet and of composite steel concrete beams" [1], the thermal behaviour of composite steel/concrete slabs was studied extensively, both experimentally and numerically. Three tests were carried out on unloaded normal weight composite concrete slabs (total height approximately 150 mm), with different type of steel decking, supported by a 200 mm wide brick wall [2]. The aim of these tests was to obtain information on the three-dimensional thermal heat flow in composite slabs near internal supports.

In order to simulate the tests a numerical model was established, consisting of eight-noded brick elements in order to model conduction in steel and concrete, and four-noded boundary elements to account for heat transfer by means of radiation and convection [3]. The model allows for exact calculation of the view-factors and the effect of melting of the sinc-layer on the steel decking.

In Figure 1, some experimental results are compared to the results of the numerical simulation. It is seen that the agreement between experimental and numerical results is satisfactory.

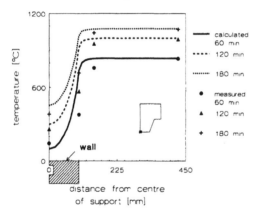

Figure 1   Numerical simulation of the three-dimensional thermal heat flow near an internal support of a composite slab.

2.2 Composite steel/concrete beam

The multi-storey building in Cardington consists of a steel frame with composite beams. The behaviour of such composite beams under fire actions has been studied in the scope of the ECSC -SA509 project [1]. As a result of that study, amongst other things, load-slip relationships were established for stud connectors at elevated temperatures.

A numerical model has been developed, comparable to the model described in [4]. In this model, slip between the steel beam and the concrete slab is modelled using truss/spring elements (Figure 2).

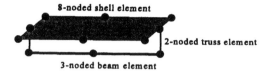

Figure 2.   Numerical model for composite beams

The model consists of three-noded beam elements, eight-noded shell elements and two-noded truss/spring elements. Physical and geometrical non-linearities are modelled, assuming measured material properties, and adopting Eurocode ENV-1994-1-2 relationships for temperature dependency [5].

The validity of the model is illustrated in Figure 3. In this figure, the results of a numerical simulation, based on measured temperatures of a fire test in a simply supported composite beam, are compared to experimental results, in terms of midspan deflection versus time. The overall behaviour is simulated quite well.

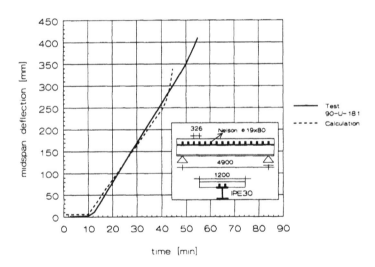

Figure 3   Comparison of numerical results versus experimental results for a simply supported composite beam.

2.3 Academic buckling problem: cantilevered beam – pin-ended strut

The academic problem consists of a rigidly connected cantilevered beam, supported by a pin-ended strut. The strut, a CHS 114 x 5 section with 1/1000 initial imperfection, is heated and the beam, a UB 305 x 102 x 38 section, remains cold. A vertical load is applied at the end of the beam (Figure 4).

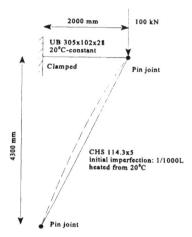

Figure 4   Academic buckling problem.

The structural behaviour was analysed with three-noded beam elements, assuming a yield stress of 345 Mpa. The results of three different numerical simulations are presented:

- geometrical linear analysis
- geometrical non-linear analysis (Total Lagrange algorithm)
- geometrical non-linear dynamic analysis

The results of the calculations are plotted in Figure 5, in terms of vertical displacement of the strut end versus strut temperature.

The results clearly indicate that, when second order effects are neglected (geometrical linear calculation) the outcome of the simulation is far too optimistic. The geometrical non-linear analysis yields more realistic results. It should be noted, however, that in order to keep the iteration process in this simulation stable, the temperature had to be increased in steps as small as 1/100°C near the top of the curve. Furthermore, when the strut buckles some of the increments did not converge, so description of the post-buckling behaviour is not reliable. The dynamic analysis showed good convergence behaviour, even near and just after buckling. Apparently, including the mass inertia has a positive effect on the stability of the numerical simulation. The post-buckling behaviour consists of a vibration around the level of static deflection of the cantilevered beam. One could expect that an even more advanced simulation, in which damping is realistically included, yields a more smooth post-buckling behaviour.

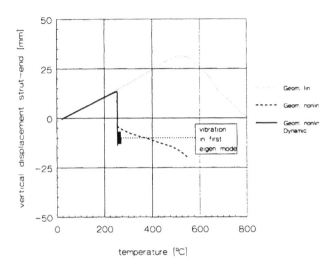

Figure 5   Analyses of buckling behaviour

It is concluded from the above exercise that a dynamic analysis may be useful in cases of sudden (local) buckling or squash failures. Including mass inertia terms and damping may significantly improve the convergence behaviour of the numerical simulation.

3 Multi-storey steel frame

3.1 Restrained beam

The first structural fire test in the scope of the Cardington project was a test on a restrained beam (Figure 6). In this figure also, the part of the structure, taken into account in the numerical modelling, is indicated. (For test details and experimental results [6, 7]). The following conditions apply:

- slab modelled with (reinforced concrete) 8-noded shell elements, stiffened with (reinforced concrete) 3-noded beam elements
- the thermal gradient in the composite slab was roughly estimated on the basis of the difference between the measured decking lower and upper flange temperatures, and the temperatures at the unexposed side [7] (Figure 7).[1]

---

1. It is noted that numerical simulation of the thermal behaviour is rather difficult as it requires data on convection coefficients and thermal characteristics of the furnace walls. In a later stage of the project, when more data will be available, these simulations will be performed.

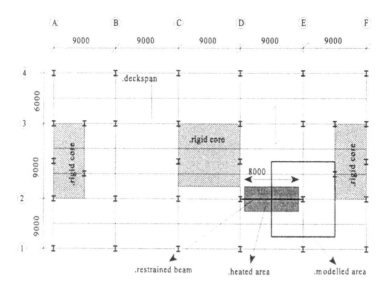

Figure 6   Numerical simulation of restrained beam test: extent of structure modelled

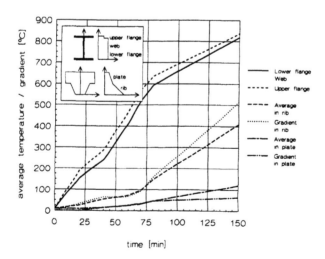

Figure 7   Adopted temperature distribution in restrained beam and heated part of composite slab.

- restrained beam modelled with 3-noded beam elements: upper and lower flange as well as the web are modelled separately which allows for the simulation of local buckling/squash failure of each member
- measured temperatures were adopted in the restrained beam [7] (Figure 7).[1]
- rigid beam to column and beam to beam connections
- measured mechanical material properties were adopted
- a secant stiffness approach was used in order to overcome bifurcation problems as a consequence of complex concrete cracking situations

The results are presented in Figure 8 in terms of maximum deflection of the restrained beam, and it can be seen that the overall behaviour is simulated quite well. After some 50 minutes, the calculated deflections fall a little behind the measured deflections; it is felt that this is due to the simplified modelling of the thermal gradients in the composite slab.

Further work will be carried out to study this in more detail, as well as the influence of other important parameters, such as the connection behaviour. It is important to note that despite the relatively high steel temperatures (800 °C and beyond) neither in the test, nor in the simulation, did failure occur.

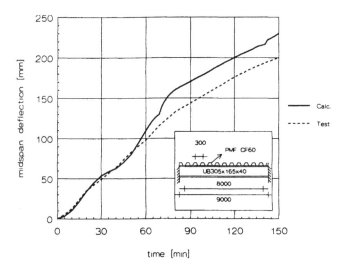

Figure 8   Numerical simulation of restrained beam test: comparison measured and calculated midspan deflections.

3.2 Plane frame

The second Cardington test concerned a plane frame (Figures 9 and 10). In these figures, the part of the structure taken into account in the numerical analysis is indicated as well. (For preliminary test results [6]).

So far, only preliminary and incomplete experimental data on the temperature distribution in the plane frame are available [6, 7].   In the calculations, this temperature distribution is schematised (Figure 10).

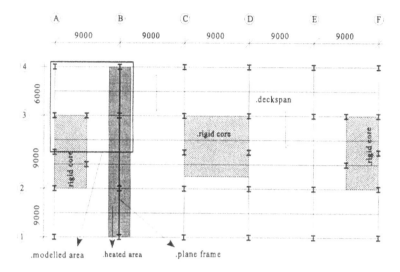

Figure 9   Numerical simulation of Plane Frame test: extent of structure modelled

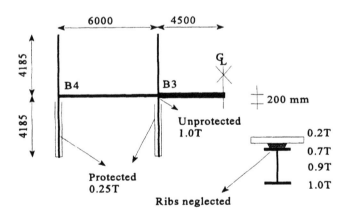

Figure 10   Adopted temperature distribution in plane frame.

The fact that the columns were protected, only up to approximately 200 mm from the level of the lower flanges of the beam (implying heated connections and upper parts of the columns) is taken into account.   Note that in view of the rough approximation of the temperatures in the plane frame, a direct comparison to experimental results is not realistic.

The numerical model is basically the same as the model used in the simulations of the restrained beam test. Three different scenarios are analysed:

• complete frame with column loadings
• complete frame, with displacement controlled movement of the node representing the connection of the main (and secondary) beam to column B3
• main and secondary beam only

The latter scenario simulates a standard fire test, which one would traditionally perform as a basis of structural fire safety assessment. Such a test does not allow to analyse the complete frame, including columns and continuous slabs. Note that the effective width of the slab needs to be estimated, and this was done by using EC4 rules [5]. The same temperature distribution was taken as in the 'complete frame' scenarios.

In Figure 11, for some calculation results related to the first scenario, vertical displacements of characteristic points of the frame are presented as a function of the lower flange temperature (T) of the main beam.

It is seen that at the level of 700 °C, column B3 squashes. This behaviour was also observed in the test, where it was found that (only the internal) column head was subjected to a sudden vertical displacement of approximately 150 mm [6]. The numerical model diverges at this point.

In the second scenario, an attempt is made to simulate the structural behaviour of the plane frame after squash failure of one of the columns. This was done by means of displacement controlled loading of the node representing the connection between column B3 and the main beam.

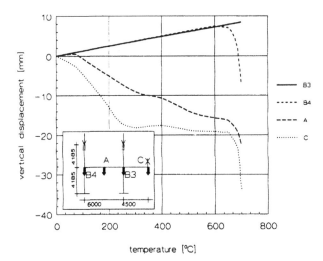

Figure 11   Numerical simulation of plane frame test: squash failure of column B3

In a first attempt, the total imposed vertical displacement due to squash failure of column B3, amounted to 15 mm. This choice is motivated as follows: if the columns were protected, as in normal practice (i.e. at least up to the lower flanges of the beams) squash in the directly exposed part of the column might have been avoided, or would have been less severe, in the sense that a smaller vertical displacement would have occurred. Under these circumstances, the squash failure displacement in column B3 is estimated roughly on 1/10 of the observed value in the test, i.e. approximately 15 mm.

The results in terms of vertical displacements as a function of the maximum steel temperature are plotted in Figure 12. It can be seen that after squash failure of column B3, an equilibrium is obtained. In fact, the frame does not fail even for steel temperatures of 1100 °C and beyond.

Evaluation of the third scenario illustrates what happens if structural restraint is neglected, and the results of this simulation are plotted in Figure 13; it can be seen that the beams fail at a maximum steel temperature level of 800 °C. For reference, the beam displacements according to the second scenario are also plotted in Figure 13. The calculations demonstrate that structural restraint, in the case of the Cardington tests, provided by the columns, secondary beams and the reinforced composite slab, may significantly enhance the fire resistance.

As soon as more experimental data becomes available, more detailed numerical simulations will be performed of the thermal response, as well as the structural response.

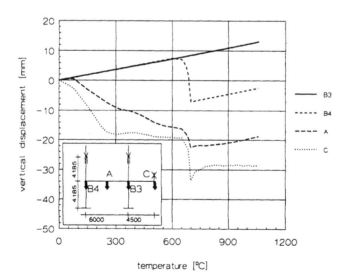

Figure 12 Numerical simulation of plane frame test: equilibrium after squash failure
of column B3

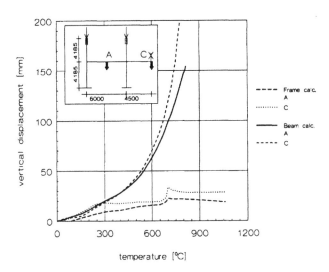

Figure 13 Numerical simulation of plane frame test: isolated beam versus complete frame

4 Conclusion

Traditionally, the structural fire behaviour is assessed on the basis of experiments on single structural elements, in which, at least in Europe, structural restraint is neglected, thus conservative results are obtained. In the ECSC research project "Behaviour of a multi-storey steel framed building subjected to natural fire effects" complete structural systems are being tested in a real building. This offers unique possibilities to analyse the effects of structural restraint. This paper describes such analyses, based on the FEM package DIANA.

Firstly, the ability of DIANA is demonstrated on the basis of fire tests on composite slabs and beams, as well as a more complex structural system. Subsequently, numerical simulations of two fire tests carried out in the scope of the Cardington project are described: the restrained beam test and the plane frame test. Adopting measured temperature distributions, the applied numerical model yields results which are in reasonable agreement with experimental findings for the restrained beam test. For the plane frame test, it is refrained from a direct comparison between test results and numerical results, since only preliminary and incomplete experimental data are available. Instead, numerical analyses are presented which clearly indicate squash failure of the tested, internal, columns. This is in agreement with observations during the test.   Furthermore, the effect of structural restraint is demonstrated by comparing numerical results of a full frame simulation to the results of a single continuous beam simulation. This effect appears to be significant and explains why in the plane frame test no global failure occurred.

5 References

1.   Kruppa, J., Zhao, B., Both, C., Twilt, L., 1994. *Fire resistance of composite concrete slabs with profiled steel sheet and of composite steel concrete beams*, Final report ECSC Agreement 7210-SA509.

2.   Both, C., Haar, P.W. van de, Twilt, L., 1994. *The thermal behaviour of three fire-exposed composite steel/concrete slabs on an insulated internal support: test report*, ECSC-Agreement 7210 SA 509, Delft University of Technology report 6.94.23/TNO-Building and Construction Research report 94-CVB-R1382.

3.   Both, C., 1995. *3D analysis of fire-exposed composite slabs*, Proceedings 3rd CIB-W14 Workshop on modelling.

4.   Zhao, B., 1994. *Modélisation numérique de poutres et portiques mixtes acier-béton avec glissements et grands deplacements resistance a l'incendie*, PhD-thesis, Institut National de Sciences Appliquées de Rennes.

5.   Eurocode 4, 1995.   ENV 1994-1-2: Eurocode 4: *Design of Composite Steel and Concrete Structures*. Part 1.2: "Structural Fire Design".

6.   Martin, D.M., 1995. *Behaviour of a multi-storey steel framed building subjected to natural fire effects*, Technical Report No. 2.

7.   Kirby, B.R., 1995.   *Behaviour of a multi-storey steel framed building subjected to natural fires*, Test 1 Restrained beam, BS doc. S423/1/PartT1.

8.   Bailey, C., Burgess, I., Plank, R., 1995.   *Computer simulation of the Cardington LBTF tests - draft preliminary report on the behaviour of the two-dimensional cross-frame fire test*, University of Sheffield.

MODELLING THE BEHAVIOUR OF STEEL-FRAMED BUILDING STRUCTURES BY COMPUTER

R. PLANK, I. BURGESS and C. BAILEY
School of Architectural Studies, University of Sheffield, UK

1 Introduction

Standard fire tests have been developed over a number of years and provide a convenient method of comparing the performance of systems against specified criteria. They have also been used to perform more investigative research into structural behaviour in fire, but for various reasons are less suitable for this purpose. Standard furnaces do not readily allow a range of typical structural geometries and loadings to be tested, and can generally accommodate only a single beam or column. The real behaviour of a structure within a fire will be significantly influenced by factors such as the rigidity of the connections, the interaction of elements and secondary structural actions. Furthermore true 'failure' may not equate to the prescribed criteria of standard tests.

The current fire test programme at Cardington will provide invaluable information about how real structures behave in fire, but is of necessity limited in its extent by the enormous cost of such experimental work. Considerable attention has therefore been directed to simulating the behaviour of structures in fire by computer analysis. Two approaches have been adopted for this. One is based on standard finite element software, as described elsewhere [1]. The second approach is to use purpose-written programs, and it is this development which is described here, together with preliminary results which have been used for designing the tests and initial validation of the software.

2 Modelling the structural behaviour of isolated elements

The high cost of even standard fire tests performed on isolated beams and columns has stimulated an interest in developing analytical methods. These were based on finite element principles, and were initially concerned with isolated beams and columns represented by a series of linear elements. The analysis is typically performed for a sequence of temperatures under constant loads, the principal variant being the material properties which deteriorate with increasing temperature. These were initially modelled using curve-fitting procedures applied to high-temperature tensile test data, using a Ramberg-Osgood formulation [2]. Other alternative representations such as the EC3 recommendations [3] have recently been added.

Because of the highly non-linear nature of material stress-strain characteristics at high temperature and the variation of temperature throughout a cross-section, subdivision of each 'element' is generally necessary. Cross-sections are therefore typically represented as a series of 'slices', each at a uniform temperature. The structural properties of the section are then determined by a process of integration, and the complete structural behaviour modelled using standard stiffness matrix procedures. This approach has been used successfully for a variety of configurations including simple steel beams supporting floor slabs on the top flange, shelf angle floor beams, and slimfloor construction [4], which differ principally in the distribution of temperature through the cross-section. For simple flexural behaviour the analysis is largely concerned with determining deflections at increasing temperature, and comparisons with standard test data have generally shown good agreement for these different forms.

Composite beams have also been studied by suitable modification of the cross-section properties. Where full interaction is assumed this is a relatively simple process, but for partial interaction a fundamentally different approach has to be adopted. This has involved modelling the beam as a series of parallel elements, representing the beam and slab separately, connected at regular intervals by elements with characteristics defined by a force-slip relationship between the steel and concrete [5].

Because of the non-linear nature of the problem, the solution procedures are highly iterative, with element stiffness dependent on an assumed level of strain and neutral axis position. Nevertheless for statically determinate conditions the solution times on a personal computer are generally very short.

Much of the work on beams was based on a secant stiffness approach [2]. This proved very efficient and computationally secure for modelling flexural behaviour and was extended to include the treatment of statically indeterminate systems such as fixed-ended and continuous beams. Results from associated studies demonstrated the importance of rotational restraint at beam supports in extending survival times for beams. This led to further refinements enabling semi-rigid connection characteristics to be included, based on assumed moment-rotation relationships at high temperature [6]. A research programme is currently being undertaken to establish appropriate connection models for fire analysis [7].

Column behaviour is largely dominated by buckling and a number of alternative approaches have been developed for analysing isolated columns. A finite strip method [8] provided a very efficient method for determining instability under axial compression, but a more traditional finite element approach, similar to that used for the beam analysis but based on a tangent stiffness formulation, was developed to give a more powerful capability [9]. This admits combined bending and compression as well as initial imperfections, and provides a complete temperature-deflection history.

A simplified method based on the Perry-Robertson approach was also found to give remarkably good results for single columns [10].

Column studies indicated that in some cases failure temperatures could be significantly lower than for beams, and it became apparent that the advantages of connection continuity for beams could not be realised if the supporting columns were to fail prematurely. This led to a progression from modelling elements in isolation to frame analysis, and subsequently complete building structures.

3 Modelling the behaviour of complete building structures

The stiffness matrix formulations developed separately for beam and column studies allowed a natural extension of the analysis to structural frames. Initially these were limited to two-dimensional models, but the importance of out-of-plane buckling quickly led to the development of full three-dimensional analysis. This was based on the same principles, but incorporated more general 'beam' elements admitting eleven degrees of freedom at each node, and accounting for geometric non-linearities as well as the material non-linearities described earlier. This is currently based on a tangent stiffness formulation and was initially concerned with bare steel skeletal frames only. However the importance of composite construction was recognised and this was introduced by modification of the beam elements, assuming an 'effective width' representation.

The effective width concept is a simplified device aimed more at design than fundamental modelling of complete structures, and the interaction of floor slab and skeletal frame is clearly better represented by the inclusion of separate slab elements. The model has therefore recently been extended by introducing 4-noded shell elements based on Mindlin/Reissner theory, and transferring the reference axis for the beam elements to a variable location, allowing more realistic connection of beam and slab [11].

Another feature which has been introduced is the effect of strain reversal from a post-elastic condition [12]. This can develop as a result of the natural progression of a fire, particularly during the cooling phase, and has involved the incorporation of material unloading paths on the stress-strain representation. In the absence of any specific guidance for steel at high temperatures, these have been based on a hysteresis loop defined by Massing's hypothesis, which is well established for normal conditions. This feature was originally developed to investigate the effect of localised and spreading fires within extensive structures, but has also been valuable in the context of the current experimental programme in modelling behaviour during the cooling phase of the tests.

4 Designing the tests

There are two principal roles for the analytical software in the research programme being conducted at Cardington, namely prediction and validation. Experimental work of this nature is very expensive and only a very small number of tests is feasible. It was therefore essential to design the test details to ensure the maximum amount of useful data. The nature of the tests, which are described elsewhere [13], was therefore carefully planned, and extensive analytical studies performed to model the expected behaviour of each test. For example, in order to design the furnace characteristics and the required heat input estimated failure temperatures must be determined. This is particularly important for cases in which timber cribs are used to ensure the availability of sufficient fuel at the start of the test. Other parameters which have been studied are described briefly below. The initial focus was on the two-dimensional cross frame test, and analyses were conducted by modelling the structure as a two- or three-dimensional skeletal incorporating composite elements with full interaction [14, 15].

4.1 Floor loading

The basic floor loading on the test frame represents the full dead load (including an allowance for raised floor, partitions, ceiling and services) plus one third of the characteristic imposed loading used in the design (2.5 kN/m^2). Additional loads could be provided by hydraulic jacks, and a number of analyses were performed for different conditions, varying from the basic load provided on the test frame up to full characteristic load. As expected the results showed that there was a reduction in failure temperature for higher load levels. For the full characteristic load the failure temperature was found to be 614 °C, and this increased to more than 700°C when only the basic load was applied. The decision to use just the basic test load was taken partly because of difficulties associated with the jacking arrangements necessary to introduce additional loads, but more importantly because it corresponded approximately to the fire loading condition specified in EC3.

The study was repeated with a 3-dimensional analysis, including the interaction between the secondary beams and the primary frame, and allowing column buckling to be modelled. A limited subframe was used, comprising the main test frame, the lift of columns immediately above, and the secondary beams supported directly on the test beams. These studies pointed up the fact that the nominal load ratio for the secondary beams is markedly higher than for the main beams, and as a result the predicted behaviour was dominated by large deflections in this part of the structure. Initial investigations conducted for cases where the load ratios in the main and secondary structure are similar suggested that the effect of the 'grillage' action could, in more general cases, be a significant factor in extending survival periods.

4.2 Fire protection

Results of the three-dimensional analyses indicated that, if fully exposed, the columns were likely to buckle before the floor beams reached a critical deflection. These analyses were therefore repeated with protected columns, assuming that their temperature was maintained at 50% of that in the beams. Although the differences in failure temperature were not large, predicting column behaviour is less reliable than for beams, and since it was recognised that columns would often need to be encased for cosmetic reasons, it was decided that for most of the tests some light protection should be provided.

Some questions were also raised about the need to fill the voids between the composite deck and the top flange of the steel beams. Analytical comparisons showed that leaving the voids unfilled had a minimal effect on survival times, and this costly activity was therefore generally excluded from the test programme.

4.3 Partial interaction

The analyses have generally been conducted assuming full interaction, although the number of connectors is strictly insufficient for this. Concerns have also been expressed that the shear forces developing in the connectors during exposure to fire could lead to stud failure and collapse. Preliminary studies conducted on single beams with both simple and rotationally restrained supports indicated that the assumption of full interaction was quite valid, and that the forces induced in the connectors during fire exposure was unlikely to cause failure. Typical comparisons are shown in Figure 1.

4.4 Connection characteristics

An important feature of frame behaviour compared with the performance of isolated elements is associated with the influence of the beam-column connections. Although some data is available for ambient temperature, little exists for fire conditions, and there are considerable uncertainties concerning the effect of the composite slab. A series of experimental and analytical studies is being conducted as part of a related project, [7] but this will not be completed for some time. It was therefore necessary to postulate moment-rotation-temperature curves based on the limited data available. Extreme upper and lower bounds were also estimated, and the frame analysed using each of these in turn. The results indicated that the behaviour was not very sensitive to the precise representation of the connection, but further investigations are continuing. In particular the action of the slab and the effect of cracking on connection rigidity are being examined.

4.5 Material models

Earlier comparisons have provided valuable guidance on appropriate methods for representing the high temperature material characteristics for steel. Concrete is a much more variable material, and little data has been recorded regarding the properties of the concrete used in the Cardington building, even at ambient temperature. A simple sensitivity study was therefore conducted, using a 2-dimensional analysis on a fixed ended beam, to compare various high temperature stress-strain models. These indicated that the results were not significantly affected either by the precise form of the stress-strain curves or the value of the characteristic concrete strength [15].

These preliminary analytical studies have proved invaluable not only in the detailed design of the tests, but also in identifying aspects of behaviour which may not have become apparent until the testing was in progress, and in establishing abort criteria. During the course of these studies, slab elements were being introduced into the model, together with modified stress-strain relationships to accommodate strain reversal. It soon became apparent that the interaction between beam and slab in fire conditions is very important, and more complex than implied by a composite element based on effective width of slab and transformed section properties. Although this more powerful modelling capability was not completed soon enough to be used in designing the early tests, it was used for predictive studies in all cases.

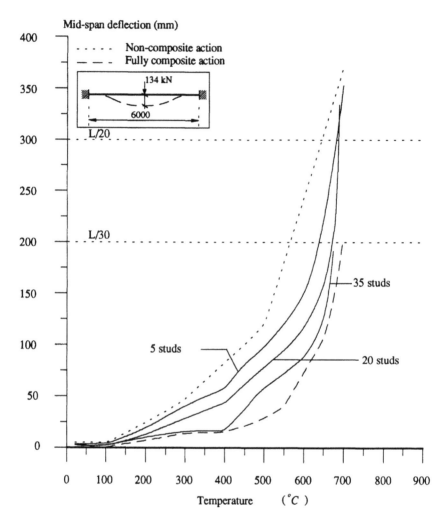

Figure 1.   The influence of the number of shear connectors on the behaviour of a
simply supported composite beam

It was also used to refine details of the later tests, and to examine the behaviour
during cooling, and it is this model which will be used during the validation
programme when the complete tests results are available. Preliminary results for
each of the tests are described below.

5 Predictive studies for validation

These studies were generally carried out before the tests, but are based on actual test details, where known. A uniform floor load of 5.48 kN/m^2 had been agreed, and the average ambient temperature yield stress of the steel determined as 308 N/mm^2 and 390 N/mm^2 for grades 43 and 50 respectively. The limiting compressive stress in the concrete, based on cube test results, was taken as 25 N/mm^2, and this was assumed to remain unaltered regardless of concrete temperature. The behaviour of concrete in tension is notoriously difficult to model, and yet is clearly an important influence when large deformations are expected. A detailed description of how this has been incorporated in the current formulation is given in ref [16] and is based on a limiting tensile stress in the concrete of 2.5 N/mm^2.

The connections at the ends of secondary beams were treated as pinned, whilst main beam-to-column connection characteristics were considered as rigid or semi-rigid. The temperature of the concrete slab and the protected columns were assumed to be 20% and 50% respectively of the hottest part of the steel beams. Since the shell elements used in the current formulation are isotropic, the ribs of the composite floor are ignored, and the slab assumed to have a uniform thickness of 70 mm.

For simplicity, the studies have been performed on subframes comprising the heated floor zone, the columns immediately above and below, and sufficient area beyond the furnace to provide a reasonable representation of the restraint afforded by the surrounding floor structure. Previous studies using a two-dimensional analysis have shown this approach to give results almost identical to those for a more complete structure [14]. Preliminary indications from the three-dimensional studies including floor slab elements confirm that the results are not very sensitive to the extent of surrounding structure included, but further investigations are currently being conducted.

5.1 Restrained beam test
In this test, a single secondary beam was heated, together with an adjacent area of floor slab (Figure 2). The analytical results are summarised in Figure 3, which also includes test deflections and comparable results for the beam modelled in isolation as a one-way spanning system. This clearly demonstrates the influence of the action of the slab in bridging over the heat-affected steelwork, with the supported beam deflections of span/20 at a temperature of about 800 °C. The analysis of the complete assembly continued to 900 °C without any indication of failure, and the test itself was terminated only because no net effective heat input to the steel could be maintained.

It can be seen that the influence of the connection rigidity is relatively small, particularly for temperatures up to about 600 °C, and with the complete structure modelled in this way, the comparison between the test and analysis is very good. However, further studies will need to be carried out when more information is available concerning realistic moment-rotation characteristics for the connection details used.

It is clear that in this case the failure temperature according to BS5950 Part 8 significantly underestimates the performance of the structure, even when modelled as an isolated beam.

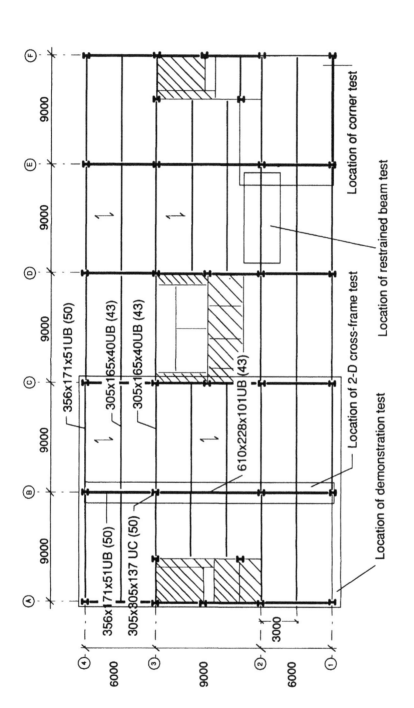

356x171x51UB (50)

305x165x40UB (43)

305x165x40UB (43)

610x228x101UB (43)

356x171x51UB (50)

305x305x137 UC (50)

Location of corner test

Location of restrained beam test

Location of 2-D cross-frame test

Location of demonstration test

Figure 2. Location of test areas on the Cardington frame

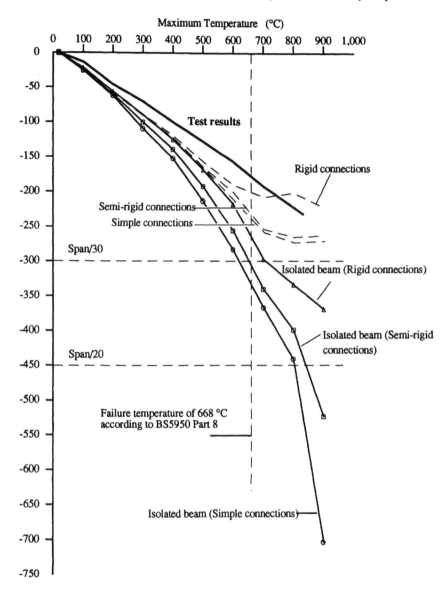

Additional mid-span displacement
due to rise in temperature (mm)

Figure 3.   Comparison of analysis and test results for restrained beam test.

The above analyses have been repeated using measured temperature profiles and the results are almost indistinguishable from those based on the simplified patterns originally assumed. This suggests that the behaviour is not sensitive to the precise representation of temperatures, and for convenience parametric studies can be conducted using simplified profiles.

5.2 The two-dimensional cross-frame test.

The two-dimensional cross-frame test involved heating the columns and beams on gridline B on one floor (Figure 2), and the analytical results are summarised in Figure 4. When the main beam-column connections are modelled as rigid, the steel temperature reaches 900 °C with no indication of structural failure, whilst the pin-ended model predicts larger deformations and a failure temperature of about 650 °C. No detailed data was available of test deflections at the time of writing, but the test was terminated at a steel temperature in excess of 800 °C, when the maximum beam deflection was about 300 mm.

5.3 The corner test

In the corner test one complete bay was heated as shown in Figure 2. The analytical results are illustrated in Figure 5 which shows the relationship between deflection of the secondary beam in the middle of the corner bay and the steel temperature. With the simplified treatment of concrete cracking, the behaviour follows the typical pattern, with gradually increasing deflection rates as the temperature rises. In this case the predicted failure temperature was 742 °C and the pattern of deflections is shown in Figure 6. Including additional areas of floor structure beyond the corner bay was found to have little effect on the behaviour, despite the additional restraint provided along the subframe boundaries.

Preliminary analyses, in which cracking and thermal strains in the concrete slab were ignored, indicated little deformation even at temperatures in excess of 900 °C. This was largely due to redistribution of loads with the slab effectively spanning longer distances.

Again, no test data was available at the time of writing, but isolated checks made during the conduct of the test indicated good general agreement, although the maximum temperature in the steel is understood to have reached almost 1000 °C, somewhat higher than indicated by the analysis.

5.4 The demonstration test

In the demonstration test two complete bays extending the full width of the building are to be heated. This represents a very severe test for the supporting steel frame since the opportunities for redistribution of load are much more limited than in previous tests. Analyses have been conducted with various levels of fire protection applied to the beams. For the unprotected condition the total deflection pattern at various critical points on the floor area is shown in Figure 7. If these are translated into net deflections, i.e. mid-span deflection of the element less the average end deflections, the behaviour of the secondary beams, which appear to be the critical steel component, is very similar to the isolated beam, with failure at about 750 °C. Similar studies with different arrangements of fire protected steelwork indicate lower deflections and higher failure temperatures, but at the time of writing the precise test details had not been finalised.

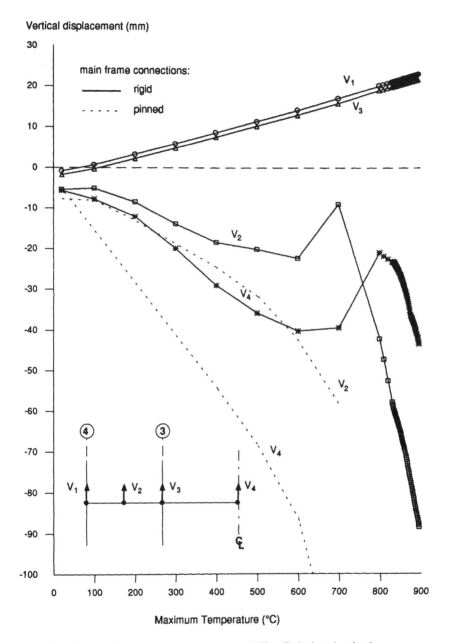

Figure 4.   Vertical displacement of beams on gridline B during the rise in temperature for the 2-D cross frame test.

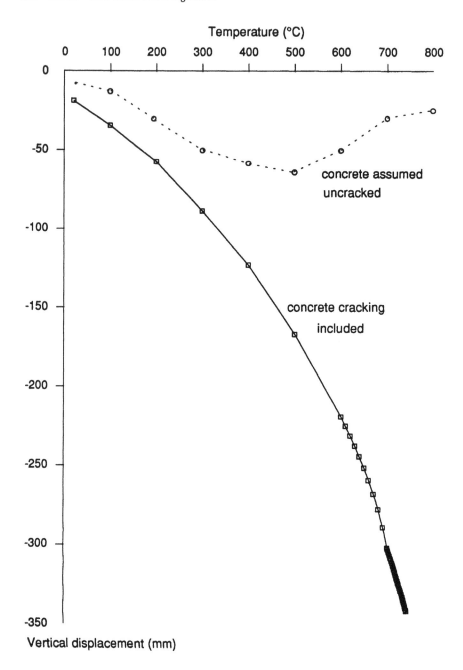

Figure 5.   Vertical displacement of secondary beam on gridline 1/2 during the rise in temperature for the corner test.

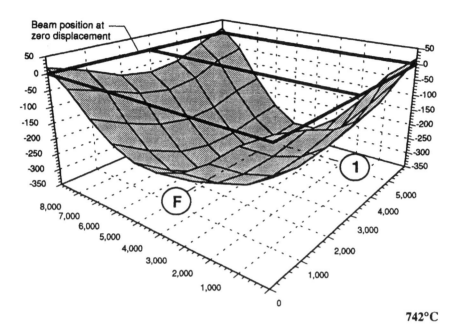

Figure 6.   Typical displacement pattern – corner test.

6 Further studies

The preliminary conclusions of the studies to date indicate that the software is capable of modelling the behaviour of steel framed building structures in fire much more accurately than methods based on skeletal frames or isolated elements. As more test data becomes available, further comparisons will be made with the analytical results, and any additional features needed to improve the modelling capability will be developed. Sensitivity studies will identify those aspects which need to be investigated in more detail, and more general cases will help interpretation of the test results in the context of more typical fire conditions.

During the course of some of the tests unexpected patterns of behaviour have been observed. Local buckling has sometimes occurred in beams close to the supports where there is significant restraint to thermal expansion, and although the current analytical formulation is not able to predict such behaviour, it is possible to model its effect on the global response of the structure. A strategy under investigation is to use standard finite element software to predict the conditions under which local buckling may occur, and to develop special modified beam elements which can then be substituted into the current stiffness matrix analysis. Preliminary studies suggest that using this approach the overall structural behaviour can be modelled very closely.

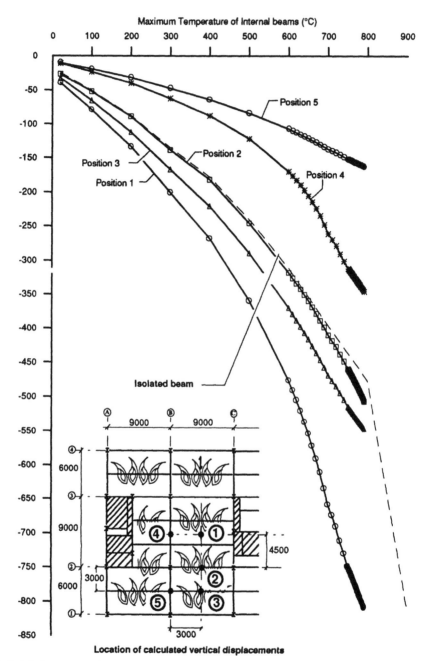

Figure 7.   Predicted deformation history for demonstration test

Another feature of the tests is the way in which the structure has behaved during cooling. It is apparent that in some cases significant axial forces are being generated. This is because of restraint to thermal expansion during the heating phase, which effectively causes the softening steelwork to shorten. As the temperature falls and the strains reverse, the material follows an elastic unloading path and tensile forces can develop, particularly in fire-affected beams. Although the main concern is the behaviour of the structure under increasing temperatures, the analysis provides for such strain reversal, and residual forces can therefore be predicted. This will allow a detailed study to be undertaken of the implications for the integrity and reinstatement of the structure.

7 References

1. O'Connor, M., *Numerical modelling of a structure subject to thermal loading*, Second Cardington Conference, March 1996, BRE Cardington

2. Burgess, I.W., El-Rimawi, J.A. and Plank, R.J., *Analysis of beams with non-uniform temperature profile due to fire*, J. Constr. Steel Research, 16 1990 pp. 169-192

3. Eurocode 3: *Design of Steel Structures*. Part 1.2: Structural Fire Design (Draft), European Committee for Standardisation, 1993

4. Burgess, I.W., El-Rimawi, J.A. and Plank, R.J., *Studies of the behaviour of steel beams in fire*, J. Constr. Steel Research, 19 1991 pp. 285-312

5. Oven, V.A., Burgess, I.W., Plank, R.J. and Abdul Wali, A.A., *An analytical model for the analysis of composite beams with partial interaction*, Computers and Structures, (in press)

6. El-Rimawi, J.A., Burgess, I.W. and Plank, R.J, *The analysis of semi-rigid frames in fire - a secant approach*, J. Constr. Steel Research, 33 1995 pp. 125-146

7. Leston-Jones, L.C., Lennon, T., Plank, R.J. and Burgess, I.W., *Elevated temperature moment-rotation tests on steelwork connections*, Proc. Inst. Civ. Eng, Structures Journal (in press)

8. Olawale, A.O. and Plank, R.J. *The collapse analysis of steel columns in fire using a finite strip method*, Int. J. Num. Methods Eng., 26 1983 pp. 2755-2764

9. Najjar, S.R. and Burgess, I.W., *A non-linear analysis for three-dimensional steel frames in fire conditions*, Engng. Struct. 18 1996 pp. 77-89

10. Najjar, S.R. and Burgess, I.W., *A simple approach to the behaviour of steel columns in fire*, J. Constr. Steel Research, 31 1994 pp. 115-134

11. Bailey, C.G., Burgess, I.W. and Plank, R.J, *The behaviour of steel framed structures subject to local fire conditions*, Proc. Nordic Steel Construction Conf. '95, Malmo, Sweden, June 1995, pp. 693-700. Swedish Institute of Steel Construction, Publication 150 Vol II.

12. Bailey, C.G., Burgess, I.W. and Plank, R.J., *Analyses of the effects of cooling and fire spread on steel-framed buildings*, J. Constr. Steel Research (in press)

13. Kirby, B., BST *European fire test programme design, construction, results*, Second Cardington Conference, March 1996, BRE Cardington

14.   El-Rimawi, J.A., Burgess, I.W. and Plank, R.J., *Model studies of composite building frame behaviour in fire*, Ottawa Conference on Fire Safety, 1994
15.   Najjar, S.R., Burgess, I.W., and Plank, R.J. *Towards rational approaches to the design of steel structures for fire resistance*, 4th Kerensky International Conference on Structural Engineering, Singapore 1994.
16.   Bailey, C.G., Burgess, I.W. and Plank, R.J., *Computer simulation of a full scale structural fire test*, The Structural Engineer (in press)

8 Acknowledgements

The analytical work described here has been performed at Sheffield University by various researchers with financial support from EPSRC, British Steel, ECSC and SCI, whose support is gratefully acknowledged.

DESIGN IMPLICATIONS OF THE CARDINGTON FIRE RESEARCH PROGRAMME

G.M. NEWMAN
Manager, Fire Engineering, Steel Construction Institute, UK

1 Introduction

The research at Cardington is being carried out by a team comprising many organisations including British Steel, BRE, Sheffield University, TNO, CTICM and The Steel Construction Institute, however the views expressed in this paper are my own and not necessarily those of the team. The paper can be divided into three parts:

- Expectations of the tests
- Findings to date
- Expectations for the future

2 Expectations of the tests

Good quality data would allow designers to have confidence in the computer programs that are emerging from universities and larger research establishments. Anyone who has tried to understand the results of a commercial fire resistance test will understand that poor data, sadly the most common, can give an air of confidence but little else. To extrapolate is extremely difficult as vital measurements are often missing. Much was expected from the Cardington tests and to judge from the many miles of wiring and megabytes of data collected, it is hoped the programme has been successful.

As well as quality data, a better understanding was required of how multi-storey steel-framed buildings perform in fire. For many years we have observed portal frames after fires and it is clear that real buildings do not perform like the individual elements in the BS476 test. Following the Broadgate fire it was astounding to see how the composite floor had behaved. In some places vertical deflections up to one metre had occurred but the horizontal deflections were negligible. The slab was playing a major part in maintaining the stability of the structure, and it was surprising to see how ductile reinforced concrete can be. It was also impressive to see the gross local deformation of the column tops and how everything was hanging together. There had been no failures. . . Was this a freak?

On the scene of the Basingstoke fire where the contents of one floor were almost totally burned out, whilst the fire had hardly scratched the structure, it was seemed that the protected steel frame had been *over* protected. Was it possible to reduce the amount of fire protection that currently applied to steel frames and still have acceptable behaviour in fire? Indeed, what is acceptable behaviour?

3 Findings to date

From a designer's point of view there is much of interest but nothing that, so far, can be directly used in the design of other buildings. The signs, however, are very good. It is clear that composite floors have a great influence on the performance of a building in fire. The experience of Broadgate has been shown to be repeatable. Also, it is apparent that, as many of us expected, the performance of a structure is better than the expected performance of the individual elements. How many beams have stood up in a fire test while at 900 °C and with no applied fire protection?

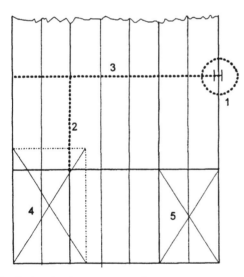

1. Isolated column tests (various locations)
2. Restrained beam test
3. Plane frame test
4. BST/ECSC corner test
5. BRE corner test

Figure 1.   Fire test locations to date

For the last five years designers have had the option of using BS5950 Part 8 and will shortly have EC3 Part 1.2 and EC4 Part 1.2. These structural design codes are largely based on the strict concept of fire resistance in that they try to predict the performance of elements of a structure in a BS476 fire resistance test. It has so far been seen in the tests at Cardington that, for beams, BS5950 Part 8 (and EC3 Part 1.2) appears to be very conservative. In the tests, beams have remained in place, supporting their load, at temperatures as much as 330 °C higher than BS5950 Part 8 would predict. Some of the beam results are summarised in Table 1. On the basis of these results designers should be thinking about different approaches other than those of BS5950 Part 8 and the Eurocodes. The challenge on this project is to devise workable engineering approaches to the problem.

Table 1.   Measured beam temperatures and BS5950 Part 8 limiting temperatures

Test beam	Flange temperature at end of test (°C)	BS5950 Part 8 limiting temp. (°C)
Restrained beam	850+	670
2-D test - 9 m span	800	720
2-D test - 6 m span	800	670
ECSC corner test - 9 m span	1000+	670
BRE corner test - 9 m span	1000+	670

Another area of interest to the designer is the effect of thermal restraint and the localised nature of damage. It was interesting to observe that the deformations caused by the corner tests were very local. – A short distance away from any of the fire compartments the structure was totally undamaged. Also, the effects of thermal restraint were confined to a limited number of members within the test area.

Concerns are often expressed about potential problems of thermal expansion where unprotected steel is used, or amounts of protection are reduced. In the tests some beams have been observed to buckle under the effects of restrained thermal expansion, but in these cases it has always been one of the beams within the test compartment, i.e. one of the beams affected by the fire. It is conceivable that, in time, we will be able to say that the effects of thermal expansion on things such as walls are small, and that concerns that walls will be pushed over by expanding unprotected steelwork are unfounded.

An important area that is under consideration is the effect of frame deformation. In any fire engineering design procedure the maintenance of compartmentation is important. It is generally accepted that the spread of fire must be controlled and that structural stability will not always be the design criterion for all elements.

3.1 Restrained beam test
The test was carried out on a 9 m span single beam, the central 7.5 m of which was heated using a prefabricated furnace. The beam was described as 'restrained' as the surrounding areas were all 'cold' and it was thought that this would effectively prevent any thermal expansion.

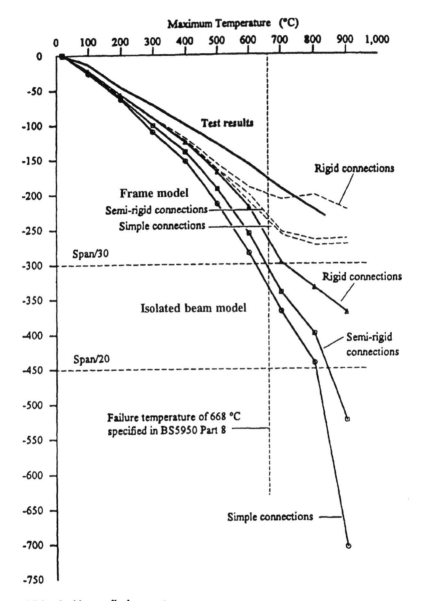

Additional mid-span displacement
due to rise in temperature (mm)

Figure 2. Comparison of test results and computer models for the restrained beam
effects are closer to the test result but it is not clear from this analysis
which of the two effects is the most important in the modelling process.

The test showed clearly two important effects that have been highlighted by the fire test programme; The effect of the floor slab, and that a beam in a frame does not behave in the same way as an isolated beam.

The central beam deflection as measured in the test, and as computed assuming various criteria, are presented in Figure 2. The computation was carried out by Dr Bailey using INSTAF. From the designers point of view the results should raise confidence as the test deflection is nearly always less than the computed deflection. It is also clear that BS5950 Part 8 and computations based on an isolated beam are very conservative and, depending on assumptions about connections, errors of the order of a factor of two are apparent. The models that include slab and connection

3.2 Two dimensional frame test

A 2.5 m wide slice was taken through the building and it was hoped to obtain information on how moments would be redistributed around the frame. An important feature of the test was that the columns were fire protected to about 300 mm below the deep main beam. The test was dominated by partial buckling of the unprotected tops of the columns which probably occurred when the bare steel temperatures were at least 700 °C. In the event the result obtained was not quite that had been expected, although a great deal was learnt about how loads can be redistributed throughout a steel frame when part of the frame loses strength.

The analysis of the results is not yet complete and so it cannot be said for certain what structural mechanisms were active. There is the possibility that the columns above the fire compartment went into tension and that the cold structure was bridging the weakened area. What is clear is that the behaviour first seen at Broadgate was repeated, and that again the floor slab played a major part. It is also reasonably certain that in buildings requiring more than 30 minutes fire resistance, columns will have to be fire protected.

Columns are vital to the overall stability of a building and although we can probably show that the loss of one column might not be disastrous, the damage caused to the building may be unacceptable to insurers and others.

Preliminary analysis has shown that the comparison between the test and computer models is quite good. The modelling has also shown that the floor slab has a major influence which is more important than the precise model assumed for the connections.

3.3 Corner tests

The general behaviour in both the corner tests was similar. The central 9 m span beam reached temperatures in excess of 1000 °C but no failure occurred, the dominant structural element being the floor slab.

There is some evidence that the outer band of the slab was in a kind of hoop compression with the central part in tension. (Figure 3). In this manner the slab can span effectively much more than that implied by simple catenary action. This type of behaviour was observed in restrained fire resistance tests on slabs carried out by UL in the USA.

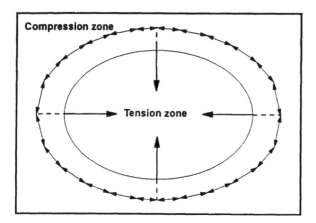

Figure 3   Possible structural mechanism in a composite floor

3.4 Column tests

The isolated column tests have been limited in scope so that structural damage did not occur. They have shown that the frame exerts little restraint in that the columns have been able to expand with little increase in applied loading.

4 Expectations for the future

4.1 The remaining fire tests

A further two large tests are planned to gain more information on the effect of fire on a multi-bay structure, and the test programme on isolated columns will be completed. In the tests so far it could be argued that the excellent behaviour has been due to the beneficial effects of the composite floor slab spanning between secondary beams 6m apart, but in the next series of tests the slab might have to span up to 9 m.

Unfortunately, fires are not confined to small bays and any eventual design recommendations cannot sensibly be confined to such limited areas.

The tests on columns, although on a much smaller scale than the large compartment tests, are also very important in formulating design recommendations and to validate design codes such as EC3 Part 1.2. In the large tests the columns will be fire protected to prevent excessive damage to the frame, but in the small tests failure will be allowed to occur. This will give us important information on such things as the effective length of columns and the limiting or critical temperature.

4.2 Future design methods

It is going to be at least two years before the results from Cardington have been fully analysed, and further studies will keep researchers busy for several more years, so can designers expect steel-framed buildings of the future look much different from today's buildings?

It is clear that any changes will be introduced very slowly, and then it is not expected that designers will use advanced non-linear thermal structural modelling techniques on each and every building, although for prestige projects this might be warranted. It is likely that recommendations, in the form of simple design rules, will first be applied to one or two buildings with research engineers working alongside consulting engineers and architects. – The marketplace is not going to change overnight. Several years ago a number of the Broadgate buildings were 'fire engineered', and at that time innovations were confined to some rationalisation of protection thicknesses using what is now contained in BS5950 Part 8 and the non-filling of voids above composite beams. Today BS5950 Part 8 is rarely used but unfilled voids are common place. The cost of protecting a steel frame is much lower than it was a few years ago owing to strong competition within the fire protection industry. This gives less opportunity for economies because it must be accepted that most innovation is led by financial considerations.

For the Cardington studies to have a significant impact, the industry must convince the specifiers and the regulators that satisfactory safety standards are maintained, and that economies of some kind can be made. One likely finding could be that existing standards for fire protecting structures are illogical. Even at this stage of the programme there is evidence that beams do not always require the same amount of fire resistance or fire protection as columns, provided that continuity through a composite floor can be guaranteed.

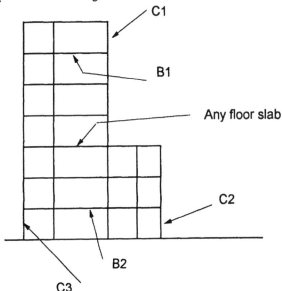

Should column C1, C2 and C3 have the same fire resistance?
Should Beam B1 have the same fire resistance as Beam B2?

Figure 4   Logic of fire resistance

The research is not a complete answer in itself but it should allow us to justify a more logical approach to structural fire safety.

The research programme being carried out by Arbed and others will be important in identifying and quantifying the effects of sprinklers and other such active systems, and going some of the way to quantify risk. Eventually we might recommend that in certain circumstances a column requires 60 minutes and the beams only 30 minutes fire resistance.

Although not directly part of the Cardington work, research has shown that some variation of required fire resistance is logical and can be justified within the height of the building. Columns at the top of a building might not require the same level of fire resistance as those towards the bottom as the consequences of column collapse at the top of a building could be considered to be less severe than at the base (Figure 4). In a few years we could have an Approved Document for steel structures incorporating more logical fire resistance requirements and the benefits of our better knowledge of structural performance in fire.

From what we have already learned at Cardington, any 'advantages' of a new approach might have to be dependent on the type of structural frame. – Sway frames may have more onerous requirements than braced frames and frames incorporating composite floors may have less onerous requirements than those with precast floor units. The use of a building might also affect the design. – Residential buildings and the accommodation parts of hotels are inherently short span structures which lend themselves to specific design approaches.

Any new design philosophy must include measures to ensure that the overall stability of the building is maintained, that the necessary compartmentation is maintained and that the building is safe for fire fighters. We face an interesting and challenging future.

PART THREE

Statics/Dynamics

CARDINGTON STATIC TESTS: RATIONALE AND PROCEDURE

R. GRANTHAM
Structural Research Engineer, Building Research Establishment, UK

1 Introduction

This paper describes the process by which a series of static load tests were designed and executed on the experimental steel-framed building at Cardington.

Much of the previous research into building performance has been carried out on elements such as beams, columns, connections and sub-frames. These last two examples start to address more complicated analysis problems of element interaction, but do not cover all the different types of interaction that occur in a building structure, such as the interaction of non-loadbearing elements. The type of test described in this paper may validate the reductionist method adopted in design and help our understanding of element interaction; hence improving the correlation of actual building performance to that predicted using finite element models.

2 Background

2.1 Test programme
The static load tests were originally conceived to take place in phases after each stage of construction on the fifth floor of the eight-storey steel-framed experimental building. The original programme was slightly modified to the following:

- Phase 1   Bare steel frame with decking (serviceability loading)
- Phase 2   Composite frame with blockwork cladding
            (A comparison with Phase 1 – 60% serviceability loading)
- Phase 3   Composite frame with blockwork cladding
            (Hydraulic jack loading to elastic limit)
- Phase 4   Tests to failure using hydraulic jack loading

2.2 Literature search

When planning such a large test series, the experience of others can often be helpful. Unfortunately, because of the impracticalities associated with testing buildings, few tests of this nature have been carried out [4]. Some guidance was found in a paper by The Institution of Structural Engineers [3] which highlighted two main criteria for the test series. First, loads should be applied over a period of more than one minute if dynamic effects are to be avoided, since one governing criteria for structural response is the rate of loading. Secondly, bedding-in-loads should be applied to the structure prior to each test to release any frictional restraints that may have been incorporated during construction. British Standards only provided guidance for acceptance testing structures or component test.

Other informative sources included published papers [1, 2, 4, 6] describing load tests on structures. These gave a qualitative view of the accuracy that may be expected from different types of instrumentation and the limitations of this data for interpretation of the building performance. For example, as part of a full scale frame test carried out at BRE [2] 23 displacement transducers were used to find the deflected profile of a 5 m long composite beam and using regression analysis, determine the beam flexural stiffness. Results showed that beam deflections could not be determined with sufficient accuracy for this type of analysis.

3 Phase 3 test series design

3.1 Method of loading

Several different loading options were considered which included water vessels, bricks, concrete blocks, lead ingots and hydraulic rams. All of these options had certain disadvantages.

The dead load options would closely simulate a uniformly distributed load at serviceability level, but because of their bulk they required almost the full floor to ceiling height at failure loads in phase 4. This was deemed impracticable. Even though hydraulic jacks had to be supported under reaction frames on the ground floor to isolate them from the structure thus avoiding unrealistic loading, the cost of this exercise was estimated to be less than the overall cost of other options. Loads could be transferred to high tensile bars passing through cores in each floor up to loading pads transferring the load to the fifth floor slab.

The pattern of loading for individual tests (four tests in phase 3) was based on a uniformly distributed load. Test 1 provided the closest approximation to uniform loading (figure 1) in which 12 pads applied load either side of two secondary beams that were being tested. Other tests applied loads directly to beams to simulate an end beam reaction that would have been present if loads had been applied uniformly to the slab only. Different elements of the fifth floor structure were investigated in each test:

- Test 1   Secondary beams connected to primary beams
- Test 2   Secondary beams connected to columns
- Test 3   9 m span primary beam
- Test 4   6 m span primary beam

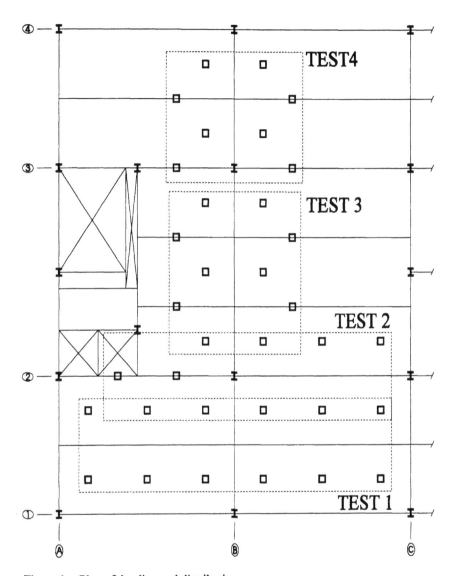

Figure 1.   Phase 3 loading pad distribution

3.2 Test area survey

To establish the condition and geometric properties of the fifth floor, surveys of the actual section sizes and material property tests were conducted. These included a slab depth survey, beam and column section size survey, concrete core compressive test to determine the concrete young's modulus, concrete cube and steel coupon tests to find the material strengths, and a survey of concrete shrinkage cracks prior to load application (figure 2).

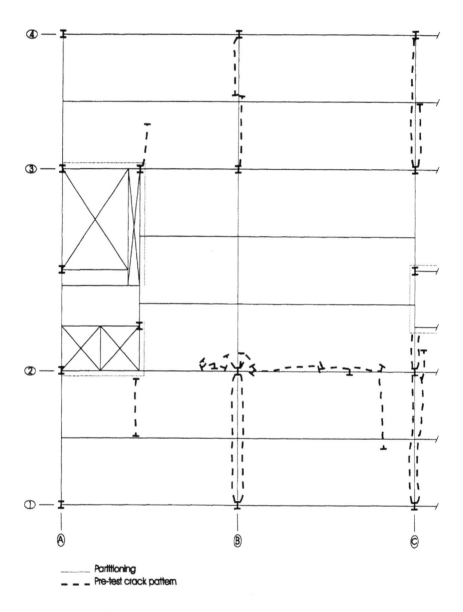

Partitioning
Pre-test crack pattern

Figure 2.   Pre-test crack pattern on the fifth floor

Table 1. Average slab thickness above beams on the fifth floor

Steel Section	Sample	Average slab thickness (mm)	Standard Deviation (mm)
305 UB	30	142.2	5.5
356 UB	10	139.3	3.2
610 UB	5	134.2	3.4

The slab thickness above all the beams to be tested in phase 3 (table 1) was much greater than the design depth of 130 mm. Average slab depths from the survey are shown in table 2 with a range of depths from 134 mm to 149 mm. Depths of 140 mm to 160 mm were typical for the slab between beam elements.

3.3 Sensitivity study

To avoid failure of the beams in these elastic tests a sensitivity study was conducted to determine a 'best estimate' of the elastic load limit for each test. This relied on data from a slab depth survey, steel sectional dimensions survey, and material property tests to determine a range for each parameter in the calculation of beam moment at yield of the bottom flange, and, thus, the maximum test load. Nominal steel section sizes were used for the calculation of test load since results of the steel sectional dimensions survey [7] showed that all measurements were within the BS EN 10034: 1993 tolerance limits.

In the absence of test data, the value for the steel Young's modulus was taken from BS 5950 with a range of ±10 kN/mm^2. The range of steel stress at yield considered, was modified to account for stresses already present in the steel following placement of the floor slab. Other values in table 2 were taken from measurements of the fifth floor, or, in the absence of survey or material test data, a nominal range was assumed.

Table 2. Variation of design parameters used in the calculation of test load

Design parameters	Range	% effect on moment at first yield	Parameters for load calculation
Steel modulus (kN/mm^2)	195 – 215	0.6	205
Steel design stress (N/mm^2)	175 – 208	18.9	208
Concrete modulus (kN/mm^2)	17.5 – 23.2	1.8	21.8
Effective width (mm)	L/4 – L/16	12.1	L/4
Slab depth (mm)	134 – 149	4.9	148
Profile depth (mm)	70, 55	0.5	70

3.4 Instrumentation

Instrumentation requirements for assessing the response of the structure to static loading was vast and amounted to almost 1000 channels of data. Six columns were strain gauged to monitor both axial force and bi-axial moment with an additional 15 columns monitoring just axial load. Over 170 strain gauges were placed on the steel beams to help determine an accurate moment distribution in each test beam. Other instrumentation included uni-axis rotation gauges, slab strain gauges and load cells.

4 Conclusion

Although enhanced understanding of element interaction and overall structural response to static load may be achieved by load testing complete buildings, the acquisition of useful data requires a much greater degree of planning in terms of instrumentation and loading system design. However, if we are to refine the conservatism still inherent in design codes this type of research must be conducted.

5 References

1.   Moss, R. M., *Guidance for engineers conducting static load tests on building structures;* Building Research Establishment, IP 2/95, 1995.
2.   Li, T., *The analysis and ductility requirements of semi-rigid composite frames*; University of Nottingham, Thesis submitted for the degree of Doctor of Philosophy, July 1994.
3.   Institution of Structural Engineers, *Load testing of structures and structural components*; Seminar on load testing, June 1989.
4.   Lloyd, R. M. And Wright, H. D.; *In situ testing of a composite floor system*, The Structural Engineer, Vol. 70, No. 12, p211 - 219, June 1992.
5.   Steel Structures Research Committee, Final report to the Department of Scientific and Industrial Research, 1936
6.   Grantham, R.; *Serviceability loading studies*, Paper presented at the first Cardington Conference, 1994
7.   Boreman, J.; *Geometric survey of steel sections on the fifth floor*, Cardington LBTF newsletter, issue 8, 1995

6 Acknowledgements

The author wishes to thank Colleagues in the Metal Structures Section of BRE and Mr J Boreman for assistance in planning and executing this series of tests.

THE RELATIONSHIP OF LIGHTWEIGHT DRYWALL PARTITIONING AND FIRE PROTECTION METHODS IN STEEL CONSTRUCTION

M.J. PRITCHARD
Manager, Technical Services Department, British Gypsum Ltd, UK

1 Introduction

The use of non loadbearing partitions comprising light gauge steel studs and plasterboard has long been a feature of construction practices in North America, and was subsequently introduced to the UK in 1965 by British Gypsum (Figure 1). They were consistently specified for commercial contracts such as hospitals and office buildings, because they were lightweight, dry and capable of providing fire resistance and acoustic separation. The real benefit was their light weight because of the effect on the overall weight of the structure.

2 System design

Development over the years has seen the introduction into the market of specialist systems to accommodate a variety of performance needs, such as Shaftwall, Staggered Stud, Separating Walls and a range of fire walls providing up to four hours fire resistance.

The one design feature which currently needs addressing is the inclusion of a detail at the head of a partition to accommodate deflection. We did not see a need for this until the mid-1980's when it was deemed necessary to include a detail in the British Gypsum literature which would accommodate up to 15 mm deflection, an arbitrary figure which was used following a survey of the typical deflections the company was being asked to design to. The detail incorporated a channel with longer legs and of a thicker gauge, and incorporated Rockwool to maintain fire resistance. This closely followed the practise adopted in North America, of cutting the board and stud short and filling the gap with Rockwool.

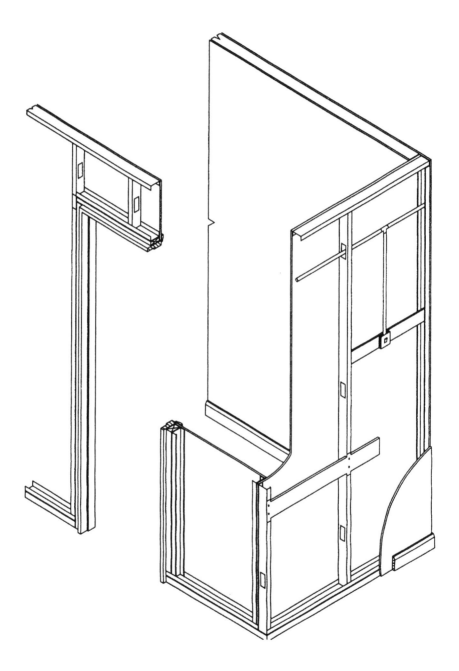

Figure 1.   Schematic of a British Gypsum partitioning system

The North Americans take the view that if partitioning is delayed for two to three months, then two-thirds of the creep deflection in a slab will have taken place. British Gypsum's view, however, was that some structural strength could be lost and the detail, whilst simple, was somewhat inflexible and structurally weak.

British Gypsum decided to adopt a `dropped soffit' deflection which required gypsum board planted at the head with the ceiling track to ensure that there was fire continuity in the form of a gypsum block, and, where appropriate, intumescent strip between the block and the soffit. This detail accommodated live deflection and fulfilled the requirement for accommodation of deflections. The question remained, how much realistic deflection occurs at the head of a non load-bearing partition wall, and how much is designed for? It must be appreciated that there is a significant cost associated with the construction of a deflection head, estimated 15 to 20% of the total partition cost, so why build something expensive if it is not necessary.

Deflection has been monitored on the eight-storey steel-framed building at the Cardington LBTF, and in a typical cell of the building 6 mm, 13 mm and 3 mm deflection was measured, giving 22 mm overall at the worst case. The design deflection was 50 mm and if detailing, then one would have to accommodate 50 mm. Is this live deflection, therefore, is it ± 50 mm or ± 25 mm. Many designers British Gypsum discuss this point with are not clear and therefore the worst and most expensive case are designed for, which, on realisation of the high cost, results in the estimate of deflection being revised. An imprecise science, and an unnecessary cost which may not suit the ethics of the Latham Report.

British Gypsum have, on the other hand, been involved in projects where deflection has occurred and not been designed for; A Bison floor had a partition built direct to its underside immediately after construction, and the resulting 19 mm deflection caused the partition to bow up to 100 mm at its mid point.

The non loadbearing plasterboard construction on the eight-storey building is Gyproc Shaftwall which was built with a nominal deflection of 15 mm. This construction is typically used in shell and core work where you are required to build from one side whilst providing fire resistance of up to 2 hours; and of course the area of the building less likely to suffer from significant deflection.

There are no significant signs of partition deformation on the building, so the deflection head of 15 mm appears to have worked, although it is quite likely that in a true commercial application a worst-case ± 50 mm might have to have been accommodated in the design. The worst deflection mid floor was recorded at 22 mm.

The corner fire test was interesting from two points of view, one was that the unprotected beam spanning 9 m deflected 224 mm in the fire, while the beam with a non-loadbearing Gyproc Shaftwall built to the underside, and, therefore, only half exposed to the fire, exhibited no appreciable deflection (a few millimetres perhaps). The other beam spanning 9 m actually twisted but the Shaftwall remained intact and the fire did not break through in any way. During the test the condition of the plasterboard wall was thought to have deteriorated, but at the end of the test the wall remained largely intact and everything outside it untouched by the fire.

The temperatures measured on the columns protected by gypsum board and with Shaftwall fixed below, and therefore half exposed, was 161 °C, and yet those unprotected were recorded at 503 °C.

3 Conclusion

There are two things to be learnt from this exercise:

1. designers must be accurate with their estimates of deflection, and whilst plasterboard may be a cheap and cheerful solution, intricate deflection details can be expensive
2. don't design for what is not necessary, and have faith in lightweight construction. It can provide excellent levels of fire protection, whilst accommodating building deflections both after construction and during a fire.

INVESTIGATION OF THE DYNAMIC CHARACTERISTICS OF THE STEEL-FRAMED BUILDING AT THE CARDINGTON LBTF

B.R. ELLIS and T. JI
Dynamic Performance Division, Building Research Establishment, UK

1 Introduction

In an earlier paper [1], presented at the First Cardington Conference, 1994, the dynamic testing of the eight-storey steel-framed building at Cardington was described, some of the initial results presented and the finite element modelling of the building discussed. The results of the testing and modelling at five stages of construction were presented in detail in a further paper [2].

This paper provides a summary of the results of both the dynamic tests and numerical modelling, and reports some of the findings from the project, discussing, in particular, three key issues. First, the non-linear behaviour which was observed in the dynamic testing and the changes of both frequency and damping with amplitude of motion. Second, the floor vibration problem which was identified in the numerical analysis and recognised as being due to some columns being horizontally offset to incorporate the building atrium. And third, the major inaccuracy in the numerical modelling, which was the representation of the infill walls, and this was recognised through a comparison of test data and numerical modelling. Several other findings and future work are also mentioned.

The building itself is described in several other papers [1, 2]. The construction has been undertaken in several well defined stages thus providing the opportunity to measure the overall characteristics at each stage. For the investigation presented here, five main stages of construction are considered:

- Stage 1   bare steel frame
- Stage 2   frame plus steel floor decks
- Stage 3   frame plus composite floors
- Stage 4   frame plus floors and walls
- Stage 5   frame plus floors, walls and static loads

2 Dynamic testing

The main objective of the dynamic tests on the steel-framed building was to determine the characteristics of the fundamental modes of vibration. Two types of test were undertaken and these are described in detail [1, 2]. The first makes use of laser measurements to monitor ambient response of the structure, i.e. its natural vibration caused by air movements within the laboratory. These measurements are processed using an FFT based algorithm to identify the frequencies of the modes of vibration. The laser has the advantage that the measurements can be made remotely and no equipment needs to be placed on the structure. The second type of test is the forced vibration test which is used to determine all of the characteristics of the fundamental modes of vibration, i.e. frequency, mode shape, stiffness and damping. This provides detailed information but requires the use of specialist test equipment attached to the building. The low amplitude measurements, taken primarily with the Laser-system, are used for comparison with calculated values. A summary of the experimental results is given in Table 1.

Table 1.   Measured frequencies at different construction stages

Stage	EW1	NS1	θ1	EW2	NS2
1a*	0.98	1.22	1.71	3.30	3.42
2**	1.31	1.55	1.67		
3	0.69	0.83	0.89	2.10	2.44
4	0.75	1.31	1.64	2.13	3.81
5	0.66	0.93	1.22	1.90	2.63

*   tested when the basic framework was erected plus the lower four steel decks
**   tested using a small vibration generator at amplitudes comparable to the laser tests

The principal axes of the building align approximately with the North-South (NS) and East-West (EW) directions, and the translation modes have been identified by these prefixes. The torsion mode has been labelled θ, with the attached number 1, referring to fundamental modes, and 2, referring to second order modes.

3 Numerical modelling

The building has been modelled and analysed using the LUSAS general purpose finite element program. The modelling is based on the engineering design and site observations. A coarse mesh with 764 nodes was adopted to model the global dynamic behaviour of the structure, and to maintain consistency in the analysis at the different construction stages, the number of nodes is constant but a different number of elements is used. The results of calculations for the six lowest frequency modes together with the building weight at each of the five stages of construction are given in Table 2.

Table 2. Calculated frequencies / mode characteristics at different construction stages

Mode order / Stage	1	2	3	4	5	6	Building weight (t)
1	1.03	1.13	1.14	1.18	1.83	1.91	325
Frame	L θ	EW B L	NS L	θ L	L2	L2	
2	1.60	1.67	1.91	3.15	3.34	3.38	373
Steel Deck	EW θ	NS	θ EW	EW2	TI	BI	
3	0.75	0.77	0.89	2.54	2.66	3.15	2302
Conc floors	EW θ	NS	θ EW	EW2	NS2	θ2	
4	0.89	1.95	2.78	2.83	4.39	4.49	2612
Walls	EW	NS	θ	EW2 θ	F	F	
5	0.70	1.55	2.22	2.27	3.37	3.38	4170
Loading	EW	NS	EW2 θ	θ	F	F	

The characteristics of the modes are described using the abbreviations as follows:

L	*Local vibration*	θ	*Rotation*
NS	*North-South direction*	TI	*Twisting in floor plane*
EW	*East-West direction*	BI	*Bending in floor plane*
F	*Floor vibration*		
2	*Second order deflection over height of building*		

4 An observation from the experimental work

From the previous sections it may be thought that the natural frequencies at any one construction stage are constant parameters. However, this is not the case, and in the previous paper [1], an example was given showing one aspect of non-linear behaviour in the fundamental EW mode which showed the frequency changing with the amplitude of motion. This is similar to behaviour noted in other structures where, with increasing amplitude of vibration, there appears to be an initial softening of the system, i.e. reducing frequencies. This is also shown by the differences seen in the results of the forced vibration test (FVT) which are at relatively large amplitudes of vibration, and the laser tests which are at very low amplitudes. The results of these two tests are given in Table 3 for the building at stage 5 of the construction.

Table 3. Frequencies measured using the two tests at stage 5 of the building construction

Test type / mode	EW1	NS1	θ1	EW2
FVT	0.62	0.80	0.96	1.84
Laser	0.66	0.93	1.22	1.90

The dynamic studies which were undertaken by TNO on the steel-framed building and reported at the First Cardington Conference [3] are also low amplitude measurements and therefore correspond to the laser measurements. The difference between BRE (FVT) and TNO measurements led to some correspondence, and the proof that these are a function of the non-linear characteristics can be shown by examining the decays of vibration obtained during the FVT.

The decay of vibration is obtained by exciting a mode of vibration at its resonance frequency, and, when a steady-state response has been obtained, stopping the excitation and recording the response as it gradually decays. An example for the EW1 mode at stage 5 is shown in figure 1.

The decay covers free vibration which is initially at high amplitude motion (2.3 mm displacement at the roof) but gradually reduces to the low amplitude levels. By dividing this decay into sections it is possible to see how the characteristics change with amplitude. This was undertaken for each of the fundamental modes, with the analysis determining both the frequency and damping by fitting a theoretical visco-elastic decay to the selected section of the curve.

One example of a section of the measured decay and the best-fit curve is shown in figure 2. The results for the fundamental modes are given in Table 4 and it can be seen that the frequency values show a gradual transition from those found in the FVT to those seen in the laser tests; it can also be seen that the damping values change with amplitude of motion.

Similar results can be seen for other buildings although the change is not as large as for the steel-framed building, perhaps because the range of amplitudes examined at Cardington was also larger.

Table 4.   Results of analysing vibration decays for the fundamental modes

EW1			NS1			θ1		
Rel. amp	Freq.	Damp.	Rel. amp	Freq.	Damp.	Rel. amp	Freq.	Damp.
1.000	0.611	2.87	1.000	0.794	3.67	1.000	0.938	4.97
0.366	0.636	1.81	0.435	0.837	3.08	0.248	1.030	4.54
0.181	0.645	1.28	0.178	0.881	2.00	0.059	1.137	3.27
0.106	0.647	1.02	0.092	0.896	1.50	0.020	1.170	2.54
0.062	0.656	0.85	–	–	–	–	–	–

An indication of what might occur for even higher amplitude motion can be gained by examining some tests conducted on a 2-storey composite frame at BRE. With this frame it was possible to apply a large range of forces with the maximum amplitude reaching 32 mm before damage occurred. The family of frequency sweeps are given in figure 3, and for comparison an equivalent sweep obtained from the steel-framed building is given in figure 4. In these tests similar characteristics to those observed at Cardington are seen at the lower vibration amplitudes, that is with increasing amplitudes the frequencies reduce and the damping increases. However, at higher levels of vibration the reduction in frequency stops and the damping appears to decrease, i.e. the apparent softening system starts to change to a hardening system at very high amplitudes just before damage occurs.

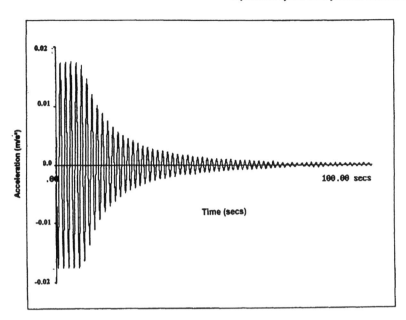

Figure 1. Decay of vibration for the EW1 mode at stage 5

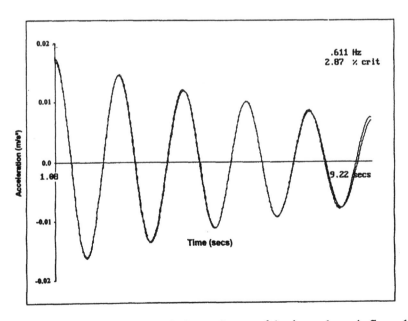

Figure 2. Measured and best-fit decays for part of the decay shown in figure 1

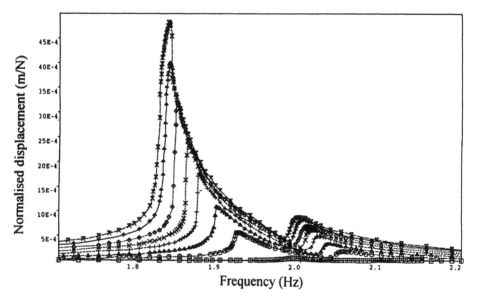

Figure 3.   Frequency sweeps on a two-storey composite frame

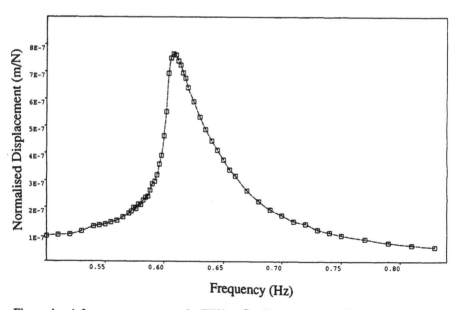

Figure 4.   A frequency sweep on the EW1 at Cardington at stage 5

These aspects of behaviour have a number of implications:

1. If an understanding of how structures fail under dynamic loading is required, it is important to consider the changes which occur leading to failure. This has widespread applications from earthquake engineering to the safety of grandstands under dynamic loading. This is not to say that this type of behaviour needs to be modelled exactly, rather that the analysts should be aware of it. Nearly all dynamic failures involve resonance, so the behaviour mentioned is relevant.
2. For engineers who use damping values in calculations, it is important to choose an appropriate damping value as the overall response of a structure at resonance is inversely proportional to damping. Thus wind engineers would be interested in typical damping values measured at amplitudes of motion similar to those with which they are concerned.

5 An observation from the numerical work

The numerical analysis was undertaken for each of the five construction stages. At stage 4 some internal modes of vibration were identified, and when examined in detail were found to be predominantly floor vibration. The floor vibrating area is directly above the entrance area and is where the columns from the first two storeys and those from storeys three to eight are offset by two metres to incorporate the atrium. The gap between the offset columns is bridged by a pair of universal beams. Although the beams have a large second moment they span 8 m and the load from the upper columns is 2 m from its support – the lower columns. The calculation indicates that the floors in this area and above had a frequency of 4.39 Hz. This mode implies that the vibration at any floor level from the second to the roof can be transmitted to other levels over the entrance area. The calculated mode shapes are given in figure 5. This suggests that the columns should be located in a straight line over the height of the building to avoid possible dynamic serviceability problems.

Following these calculations the authors experienced a disturbing vibration on the upper storey caused by movement of a loaded trolley on the fifth floor over the entrance area. A test was conducted to measure the fundamental floor frequency which was found to be approximately 7 Hz.

It is evident that there is a large difference between the calculated and measured frequencies of the floor vibration. This is to be expected when the floor areas within the columns of the framework are only represented by two or three elements, which is reasonable for examining the overall translational modes of the building, but too crude for assessing floor vibration. If the floor vibrations were to be modelled at the appropriate level, it would require many more elements per floor and also the eccentricities of the floor and beams would need to be incorporated. In addition, there are potential errors in the input data both for Young's Modulus and the cross-section of the floor. However the original model did identify the potential problem even though the frequency wasn't predicted accurately.

The floor serviceability problem could be cured by erecting infill walls between the first and second floors below the offset column, between the central column and the edge column of the south face of the building.

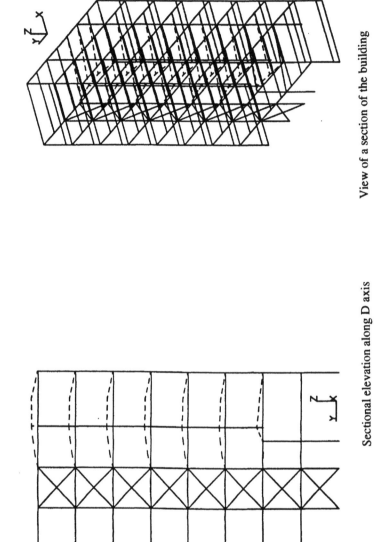

Sectional elevation along D axis

View of a section of the building

Figure 5.   Mode shape for the floor vibration

The walls would act as very deep beams and stiffen the area significantly. Calculations show that this would remove the vibration problem. From the numerical work the following conclusions were drawn:

- Difficulties were encountered for modelling the steel decking and walls
- The modelling of the building from construction to completion identifies the effect and contribution from the main structural elements, i.e. floors and walls
- Parametric studies help to identify the degree of modelling error
- The FE modelling reveals the buildings dynamic behaviour, some aspects of which can be anticipated from the structural layout of the building, e.g. the asymmetry in the building leads to coupled translation and torsion in some modes.

6 An observation from the joint experimental and numerical work

In this paper both experimental and numerical work are discussed. With the measurements the information can be considered to be accurate but incomplete (i.e. only a few modes are examined). On the other hand the numerical study gives complete but inaccurate results. Thus the experimental and numerical studies are complementary and their combined usage is sensible.

One way of using the experimental results would have been to refine the numerical model at each stage, however the model presented has not been improved using the experimental data, but instead provides an example of both typical and accumulated errors.

The comparison between both experimental measurements and numerical results can lead to other findings. To give one example, consider the major source of inaccuracy in the numerical model, which is modelling the infill walls. At stage 4 the walls were erected and they have a dominant effect on the overall stiffness of the structure and are, in fact, important for any framed structure. The blockwork walls have only been erected around the outer face of the building, with full-height walls on the East and West faces and dado height walls (900 mm) on the North and South faces. Both the measured and calculated frequencies before and after the erection of the walls are shown in Table 5.

Table 5. The changes due to the addition of infill walls on both the measured and calculated frequencies

	EW1		NS1		θ1	
	Meas.	Calc.	Meas.	Calc.	Meas.	Calc.
Stage 3	0.69Hz	0.75Hz	0.83Hz	0.77Hz	0.89Hz	0.89Hz
Stage 4	0.75Hz	0.89Hz	1.31Hz	1.95Hz	1.64Hz	2.78Hz
% Increase	9%	19%	58%	153%	84%	3.12%

By comparing measurements and calculation it was found that the increase in calculated frequencies is bigger than the increase in measured values. Also the increase in NS and torsion frequencies is far larger than for the EW modes. This shows the walls in the calculation are significantly stiffer than they should be, especially for the full-height walls on the east and west faces. This provides clear and useful feedback for identifying errors in the models of the walls.

The walls at the two ends are actually divided into several areas by beams and columns, and the construction gaps between the walls and columns and beams are unavoidable. However, in the numerical modelling it is difficult to represent the real situation accurately and the assumption of continuity between walls leads to significant differences between measurements and calculations. This data provides an opportunity to refine the numerical model and also evaluate various methods for modelling infills which have been proposed world-wide.

The combined experimental and theoretical work also served to show the contribution of the composite floors to the overall translational stiffness of the building and the effects of loading on stiffness. It is disappointing that it was not possible to test the bare steel frame (stage 1) due to the way the building was constructed, as this would enable an assessment of the accuracy of the model of the bare frame to be made with its assumed boundary and joint conditions. It would also have enabled the errors due the modelling of the steel decking to be determined.

7 Conclusion

The steel-framed building at the Cardington LBTF has provided opportunities for a number of different research programmes and here some of the results of one investigation on the building have been considered.

The chance to recognise where errors occur in numerical modelling, through comparison with measurements at various stages throughout the construction, is perhaps unique and focuses on an important area of engineering analysis. This type of study provides a means for improvements in engineering design which is likely to become of increasing importance as computer analysis is used increasingly in engineering.

Now the construction stages of the steel-framed building are complete, tests which lead to local damage are being undertaken. These local tests are not relevant to the study of the overall building characteristics as discussed in this paper, although for the final tests, which produce global damage, frequency measurements will be made.

This study helps to achieve a better understanding of one particular building, and identifies weakness in the numerical modelling and design, although by itself the study does not provide general guidance for engineers. A similar type of study was undertaken considering the performance of concrete large panel structures [4], using dynamic measurements taken by BRE, although in this case only measurements on finished buildings were available.

It is anticipated that similar studies will be undertaken on future buildings at Cardington where the staged construction can be studied in detail. It is through such studies, linking mathematical models with actual performance, that the accuracy and weaknesses in modelling whole buildings will be revealed and thus the opportunity to learn and improve designs.

8 References

1.   B R Ellis. *Dynamic testing.* The First Cardington Conference, Cardington 1994.
2.   B R Ellis and T Ji. *Dynamic testing and numerical modelling of the Cardington steel-framed building from construction to completion.* Offered for publication in the Structural Engineer, 1996.
3.   B J Daniels. *Serviceability testing at Cardington.* The First Cardington Conference, Cardington 1994.
4.   Milne, J. S. *Modelling overall building behaviour for design purposes.* PhD thesis, Heriot-Watt University, Feb. 1992.

THE USE OF HIGH EXPLOSIVE CHARGES WITHIN THE BRE CARDINGTON LABORATORY – AN EXPERIMENTAL FEASIBILITY STUDY

D. CROWHURST and S.A. COLWELL
Fire Research Station, Building Research Establishment, UK
MAJOR J. MACKENZIE
DRA Chertsey, UK

1 Introduction

The bomb continues to remain a traditional choice of weapon for the terrorist. Internationally, bombings have accounted for nearly half of all terrorist attacks since 1968 [1]. The threat ranges from the large vehicle bomb to the small package or letter bomb.

Following the construction of the Large Building Test Facility (LBTF) within the Building Research Establishment (BRE) Cardington Laboratory, there has been a great deal of interest in its potential use for studies involving the use of high explosives. From the outset it was appreciated that such experiments could cause damage to the internal Laboratory facilities, and possibly to the external fabric itself, and therefore a limit on the size of explosive charges used would have to be imposed. It was also recognised that if the charge weight was (for safety or other operational reasons) set too low, then the practical use of Cardington and the LBTF for high explosive work would be restricted.

Estimates of the maximum charge weights which might be used within the confines of the hangar ranged from 2 kg to 4 kg TNT equivalent. Due to the complex geometry of the laboratory environment, the uncertainty in these estimates was very high.

It was therefore decided that the most effective means of assessing the potential use of high explosives within the Cardington Laboratory would be to carry out an experimental feasibility study. The aims of the trial were to assess, for increasing charge weights:

- the collateral damage levels within the hangar structure
- the response of the external hangar fabric
- other factors such as noise and disturbance in the neighbourhood of the laboratory

• whether suitable data measurements for modelling applications could be obtained

This paper presents the findings from this feasibility study and compares some of the preliminary results with data obtained from two predictive methods.

2 Experimental work

2.1 Test arrangement
The experimental programme involved the free air discharge of PE4 high explosive in the open area of the Cardington Laboratory, in front of the LBTF. Figure 1 shows the location of the discharge point, 15 m above the hangar floor level.   Table 1 summarises the experimental conditions used for this study.

Table 1.   Summary of Experimental Conditions

Test	Number of L2A1 Detonators	Charge Weight (g PE4)
1	1	500
2	1	1000
3	1	1000
4	1	1500
5	1	2500
6	1	2500
7	4	880

3 Measurements

3.1 Pressure and displacement measurements
Figure 1 shows the location of the monitoring equipment in the immediate vicinity of the charge site, together with the floor plan of the fixed facilities within the hangar. Table 2 summarises the sensor types used at each monitoring location.   Pressure measurements were made using Kistler piezo-electric quartz pressure transducers, and the deflection of the structures were monitored using RDP Electronics DCT type displacement transducers.

3.2 Data recording
The experimental data was recorded on two Data Laboratories DL6000 Multitrap Modular Waveform transient recorders, triggered using a single geophone located approximately 15 m from the discharge point.

3.3 Observations
Prior to the experimental programme, a full video and photographic record of the most vulnerable facilities within the hangar structure was made, and were used as a

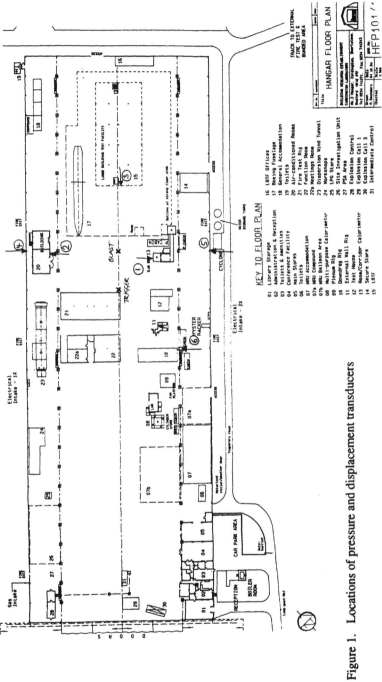

Figure 1. Locations of pressure and displacement transducers

baseline for any damage subsequently observed. Further photographic records were made of damage observed during the experimental programme.

Table 2.   Transducer device and location – local to the charge

Transducer identity	Location
DT1 & PT1	8' x 8' masonry wall
DT2 & PT2	Conditioning room breeze-block wall
DT3 & PT3	LBTF external breeze-block wall, 4th floor
DT4 & PT4	North low-level external cladding
DT5 & PT5	South low-level external cladding
DT6 & PT6	South high-level external cladding

DT corresponds to a displacement transducer
PT corresponds to a pressure transducer

4 Results

4.1 Pressure and displacement
Table 3 summarises the maximum and minimum pressure and displacement recorded on the external cladding for each test (DT4 & DT5). Figure 2 illustrates a typical pressure and displacement time history recorded on the external cladding from Test 5, 2.5 kg PE4.

4.2 Damage survey
Table 4 details damage sustained by the structures within the hangar upon completion of the experimental programme.

5 Data analysis

5.1 General
Triggering the data recorders using the single geophone proved to be an extremely reliable technique. However, because it was located 15 m from the charge it did mean that zero time had to be obtained by adding the calculated time of arrival at the geophone to each record. This introduced a potential source of error, albeit a constant one, in the determination of the time of arrival of the blast wave.

5.2 Predictions
The prediction of overpressures, and the resulting impulses from high explosive events within a containment structure is not straightforward. On a simplistic level, the PC-based programme CONWEP, written and distributed by the US Army Corps of Engineers Waterways Experiment Station (WES), gives a well established and simple method of predicting the pressures and impulses on a structure or structural element. This applies for a range of explosives at varying distances. A limitation of CONWEP is that it does not consider PE4, possibly the most widely used military

Test	Charge Weight (g PE4)	North Low Level Cladding								South Low Level Cladding							
		Pressure (kPa)		Time (ms)		Displacement (mm)		Time (ms)		Pressure (kPa)		Time (ms)		Displacement (mm)		Time (ms)	
		Max	Min	Max	Min	Max	Min	Max	Min	Max	Min	Max	Min	Max	Min	Max	Min
1	500	No Data Record				4.9	-4.4	345.3	386.6	No Data Record				4.0	-3.2	72.0	220.7
2	1000	No Data Record								No Data Record							
3	1000	No Data Record				9.3	-7.6	336.4	378.4	No Data Record				7.3	-5.8	64.1	213.3
4	1500	4.5	-2.0	50.0	64.6	12.2	-10.5	335.6	377.8	4.0	-1.5	46.9	63.5	9.5	-7.2	62.5	211.6
5	2500	4.4	-2.0	52.0	63.2	17.8	-14.4	338.0	379.1	4.0	-1.1	47.9	58.1	15.2	-10.0	64.1	212.5
6	2500	4.2	-2.1	52.2	63.5	17.6	-13.7	339.2	381.0	5.7	-1.8	48.6	65.9	14.4	-10.7	65.2	213.2
7	880	2.4	-2.0	37.7	46.4	8.0	-6.1	52.0	381.0	3.2	-1.5	33.3	42.9	6.4	-5.0	48.6	213.2

Time = 0 at the Trigger point, 15m from the Blast Site

Table 3   Summary of Low Level Cladding data

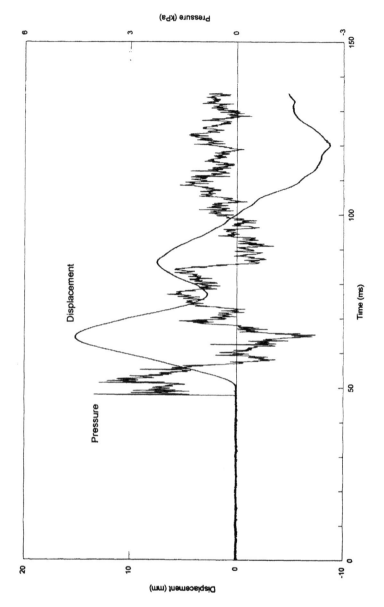

Figure 2.   Pressure and displacement time histories – Test 5, South low-level cladding

Table 4.   Damage report made after completion of experimental programme (Fig.3)

Location Ref	Ref	Damage
21	Single-storey, steel-frame structure, superlux infill panels on timber frame.	Superlux infill panels cracked.
22	Single-storey, steel-framed structure, infill panels of sterling board on a timber frame.	Clock damaged. Fluorescent light tube smashed/fittings dislodged.
12 (House)	Garage	Side entrance door and frame damaged.
	Internal damage	Window latches damaged and front door frame moved.
	External damage	Roof tile broken and debris from hangar deposited on roof.
Caravan No.3	Wood panels with single-glazed windows.	Fluorescent light tubes smashed. Windows cracked.
Portakabin 113	Standard Portakabin const. Single-glazed.	All fluorescent light fittings dislodged.
13	Masonry/blockwork structure.	Wooden doors broken. Fluorescent light fitting dislodged and smashed. Blanket fire insulation removed from rig door. Blockwork on annex cracked.
8	Masonry/blockwork structure. Single-glazed windows/doors.	Glass window pane broken. Glass pane displaced/misshapen.
Room/corridor portakabin	Standard Portakabin construction, Single-glazed.	Glass window pane broken. Fluorescent light fittings dislodged.
9	Single-storey blockwork structure, fitted with suspended ceiling.	Suspended ceiling tiles cracked and displaced.
11	Four-storey blockwork structure of complex geometry with associated light fittings. Control room: Single skin caravan with glazed windows.	Two lamps broken. Glass window pane broken in caravan.
20	Single-storey blockwork building.	Expansion of existing crack damage in the blockwork.

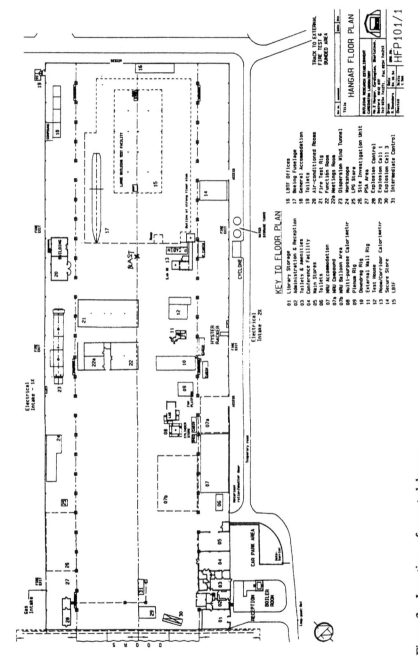

Figure 3. Locations of reported damage

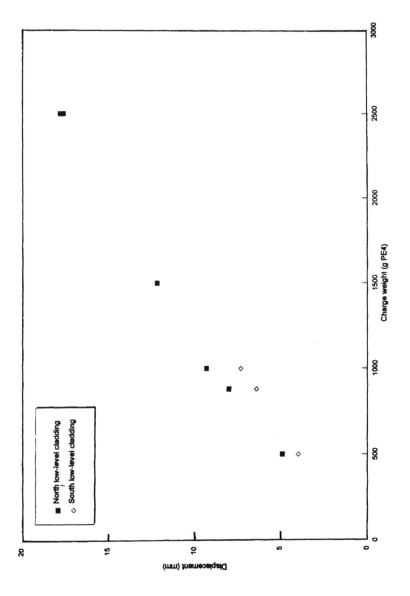

Figure 4. Pressure and displacement time histories – Test 5, South low-level cladding

explosive for demolition in the UK, although C4 data is available. C4 is chemically the same as PE4 in terms of the actual explosive, but contains a different binder. A further limitation of CONWEP is that although it has algorithms to cope with the phenomenon of clearing, it does not predict the multiple reflections resulting from an explosion in a confined space.

WES are continuing to develop BLASTX, a PC-based ray tracing programme which allows prediction of the pressures and impulses resulting from internal explosions, but it has not been widely distributed outside the US. It could possibly prove the ideal vehicle for the prediction of blast phenomena within a complex structure like the Cardington laboratory.

Probably the most appropriate predictive tools available are the hydrocodes. Many codes exist, but fortifications work at Defence Research Agency (DRA) Chertsey, has concentrated on the Finite Difference code AUTODYN. Hydrocodes are expensive, time consuming to use and require a more experienced user than other codes such as CONWEP and BLASTX.

5.3 Panel deformation data

The prediction of displacement is more difficult than the prediction of either the pressure or impulse duration. Figure 4 shows a plot of displacement against charge weight for transducers DT4 and DT5. The lack of permanent panel deformation, even at the largest charge size, showed that the panels were responding below their elastic limit. This allows the upper charge weight limit for the hangar to be set with a high degree of confidence. At this stage it was not considered practical or cost effective to model panel deformation, although this would be possible using the hydrocodes.

5.4 CONWEP Predictions

Only a portion of the data could be analysed before this conference. CONWEP was used to validate the pressure readings obtained from transducers PT1 to PT6 for Tests 4 to 7. Simplifying assumptions were made in order to make the analysis easier. It was assumed that the explosive used was C4, and that the incident wave form was normal to the target, which it was not. Since an angle of incidence of less than 90° will reduce the reflected pressure, predictions higher than the actual measurements recorded would be expected.

Figure 5 shows a typical comparison, using Test 4, of the actual and predicted arrival times for the blast wave. CONWEP typically over-predicted the reflected pressures by approximately 20% and predicted a time of arrival of the blast wave approximately 16% later than measured.

5.5 AUTODYN predictions

AUTODYN 2D was used to predict the reflected pressures on the hangar sheeting at the locations of transducers PT4 and PT5 for the 2.5 kg air burst (Tests 5 and 6). The 2D version of AUTODYN was chosen because the transducer positions were diametrically opposite the site of the explosion, and Test 5 and 6 were identical and therefore comparable. C4 properties were used to model PE4.

Whilst the external geometry of the Cardington hangar is relatively simple, the plethora of internal structures led to a congested and complex geometry. The initial

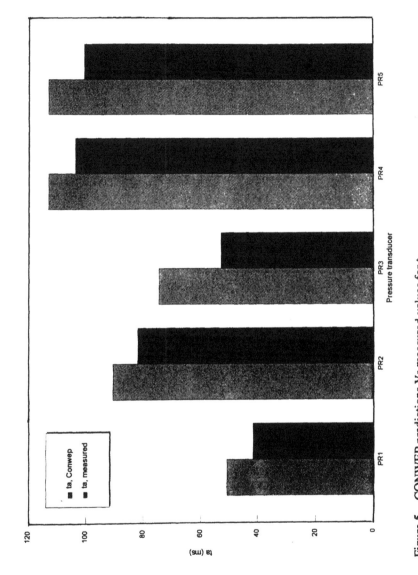

Figure 5. CONWEP predictions Vs measured values for t_a

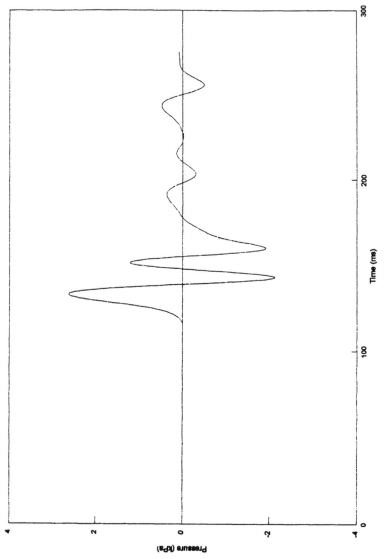

Figure 6. AUTODYN 2D plot – 2.5kg PE4 at 15m height

study considered the internal space as a parallelepiped of the correct width. A 2D simplification was then adopted to reduce run time. This over simplification was considered valid because of the relatively large height and length compared with the width. Although it would theoretically be possible to model the geometry of internal structures using a 3D hydrocode, the effort required would prove prohibitive.

The floor and hangar wall were modelled as rigid boundaries, and although material models for both steel and concrete exist, the extra effort involved in their use would probably add little to the accuracy of the final solution.

The AUTODYN plot is shown in Figure 6. AUTODYN was found to under-predict the reflected pressure and predict a later blast wave time of arrival.

5.6 Comparison of predictions

Predictions for transducers PT4 and PT5 for Tests 5 and 6 were made using CONWEP and AUTODYN. These are compared in Table 5 below.

Table 5. Comparison of Measured and Prediction Explosion Pressures

	Time of arrival, ta (ms)	Pressure (kPa)	
		Pmax	Pmin
Measured (Average of 4 readings)	102	5	-2
Calculated CONWEP	110	6	–
Calculated AUTODYN	116	3	-2

Despite the simplifying assumptions, the predictive methods used gave reasonably accurate solutions. CONWEP over-predicted and AUTODYN under-predicted reflected pressures. Both predicted a later time of arrival, possibly because of the C4 assumption. AUTODYN accurately predicted the negative phase pressure (CONWEP cannot predict negative pressures).

6 Conclusion

Air blasts up to 2.5 kg in weight of PE4 high explosive (3.4 kg TNT equivalent) can be accommodated within the BRE Cardington Laboratory without immediate threat of damage from pressure effects on the external cladding. For charges up to 2.5 kg PE4, limited collateral damage to windows and other structures within the main fabric of the Laboratory will occur. The most significant damage observed was the extension of existing cracks in damaged masonry. Other damage was minor (i.e. non-structural) including the breaking of unprotected windows, localised damage to door and window fittings and damage to light and other unsecured fittings.

Displacement measurements indicate that there may be potential to use higher charge weights without immediate threat to the main hangar fabric, but the use of larger charge weights could increase the level of damage to the internal fabric and test rigs within the hangar structure.

Predictions based on CONWEP and AUTODYN 2D were found to give reasonably accurate solutions for the air blast scenarios.

7 References

1.   *Protecting Buildings from Bomb Damage.* USA National Research Council. ISBN 0-309-05375-7. National Academy Press 1995.

8 Acknowledgements

The authors wish to acknowledge the assistance received from C Cowdery, DRA Chertsey and R Kennedy, DRA Fort Halsted in the undertaking of this feasibility study.

PRECISION IMPACT TESTING AND ITS APPLICATION FOR STRUCTURAL DYNAMICS RESEARCH

T. KRAUTHAMMER
Professor of Civil Engineering, The Pennsylvania State University, USA
D.B. MOORE
Structural Design Division, Building Research Establishment, UK

1 Introduction

One of the last frontiers in structural dynamics is the behaviour of buildings under the effects of impact and explosions. Traditionally, this area has been under study by the military fortifications people, who performed most of the research by using field testing with high explosives (HE), however, in recent years the fortification sciences community has expressed a deep concern with both the high cost and the well known lack of precision associated with field HE tests. Although valuable for many purposes, data from field tests may not be useful for precise description of complicated structural behaviour, or in support of development, verification and validation of numerical capabilities. The anticipated continuing decline in funding for R&D in fortification science and technology is particularly critical when large-scale tests are considered.

Due to current limitations in numerical simulation capabilities, R&D organisations must rely on expensive tests to provide recommendations on specific problems. If the cost for obtaining such answers cannot be dramatically reduced, it could have a serious adverse effect on the viability of R&D in this critical field of science and technology. This paper examines a promising approach that shows significant potential to overcome this problem, and provides recommendations on how to implement it.

Similarly to requirements in other scientific areas, modern fortification technologies are founded on a combination of precision tests and sophisticated numerical simulations. The linkage between these two essential components has increased gradually over the last half century, and the driving force behind both the capabilities and requirements has been the rapid evolution in computer power. Nevertheless, as a result of current enhanced capabilities in both experimental and numerical analysis, the needs for stronger interaction and collaboration between researchers in these areas is greater than ever. This point of view has been emphasised in three related events. The Norwegian Defence Construction Service (NDCS) sponsored a three-day workshop [1] during which the invited participants discussed and evaluated the current knowledge and requirements for future research, with respect to structural concrete slabs subjected to modern weapon effects. About three months later, the Defence Nuclear Agency (DNA) sponsored a two-day workshop devoted to verification and validation of non-linear structural dynamics codes. Again, in October 1995, the DNA sponsored a one-day conference on verification and validation of non-linear structural dynamics codes. Although only very few individuals participated in all three events, the main conclusions and recommendations were essentially identical. It is clearly recognised that numerical simulations will play an increasingly more important role, eventually replacing many experimental studies. This shift is expected to translate into very significant cost reductions in future (most possibly before the end of this decade) fortification related R&D. However, to enable this transition and ensure that it will be effective, it is required to concentrate in the near term on precision testing, as an integral part of development, verification and validation of the eventual numerical capabilities.

Unfortunately, most R&D organisations active in the fortification area have not developed precision testing capabilities that could be used for the study of both small and full-scale structural components. Such facilities are essential for obtaining test data that could be used to gain a deeper understanding of medium-structure interaction (MSI) and/or structural response to correlate the well defined and reproducible loads with carefully measured corresponding responses. This is not surprising as most R&D organisations have been heavily involved in mission-oriented work. They have traditionally responded to their customers' needs, and did not have the time and resources to develop innovative technologies in anticipation of conditions that were likely to exist in the distant future. These conditions, however, were anticipated by a few individuals who embarked on a gradual and systematic development of innovative approaches that could be used effectively in support of computer code validation and verification.

Researchers have carefully considered various experimental approaches that could be adopted for this collaborative precision testing activity. Based on the available information, and following discussions during the NDCS-sponsored workshop on structural concrete slabs [1], it has been decided to adopt impact testing approaches for achieving the stated objectives. The reasons for this selection are summarised below:

1. Impact testing is among the very few experimental techniques in short duration dynamics that ensure a precise delivery of energy and impulse to a test article. This general approach has been used for many years in material testing (e.g. fracture toughness evaluation, split Hopkinson bar experiments etc.). Clearly, identical amounts of energy and impulse can be ensured in multiple tests but this is very difficult to achieve with other energetic testing methods. Therefore, this approach is definitely a "precision testing" approach.

2. Experimental data from impact testing can be used for code validation or verification. This is true for any type of load obtained in the test, even if the load has no direct relationship with a specific weapon effect. Nevertheless, one can

achieve significant control on the important parameters which define a load function: Rise time, peak load, duration, and shapes of the loading and unloading branches of the function, as discussed below. It is possible to obtain load functions that are generally similar to those obtained from typical weapons, however, a direct relationship between such pulses has to be established and verified.

3. The combination of drop height, mass, interface material and impactor geometry can be used to obtain simulations of the positive loading phase of various weapon effects (with emphasis on conventional weapons). Although the lack of a negative loading phase is an inherent deficiency of the approach, similar problems exist with many other simulators. Nevertheless, the advantages listed in 1. and 2. above seem to outweigh this problem.

2 Some aspects of impact testing

The problem of impact between two bodies has been studied extensively [2, 3, 4, 5, 6, 7], and it has been shown that the impact problem can be formulated in light of Newton's Second Law of motion, as briefly outlined by Krauthammer during the First Cardington Conference 1994:

$$F = M\ddot{u} \tag{1}$$

Now, consider a mass, M, impacting a structure with resistance, R(u), and derive the equation of dynamic equilibrium:

$$M\ddot{u} - R(u) = 0 \tag{2}$$

but the structure also has a mass, and there is an impact resistance between the mass and the structure. As a result one needs to rewrite the equation of equilibrium:

$$M_1\ddot{u}_1 + R_1 (u_1 - u_2) = 0 \tag{3}$$

$$M_2\ddot{u}_2 - R_1 (u_1 - u_2) + R_2 (u_2) = 0 \tag{4}$$

where M1, $\ddot{u}_1$, u_1 are the mass, acceleration and displacement of the impacting body (impactor), respectively. M_2, $\ddot{u}_2$, u_2 are the mass, acceleration and displacement of the impacted structure, respectively. R_1 and R_2 are the impact and structural resistance, respectively.

This system of equations describes the case of "Hard Impact" where the equations of dynamic equilibrium for the structure and impacting body are coupled. In many cases the displacement of the impacting mass is much larger than the structural displacement (i.e. $u_1 >> u_2$), and therefore, Equation (3) can be rewritten as:

$$M_1\ddot{u}_1 + R_1 (u_1) = 0 \tag{5}$$

Equation (5) can be solved together with Equation (1) to give:

$$R_1(t) = F(t) \tag{6}$$

Now, Equation (4) can be rewritten as follows:

$$M_2\ddot{u}_2 + R_2(u_2) = R_1(u_1) = F(t) \tag{7}$$

This case, where $u_1 \gg u_2$, permits one to uncouple Equations (3) and (4), and it is defined as "soft impact". One can calculate the impact forcing function, $F(t)$, from Equation (6) by assuming that the responding structure is rigid (i.e. $u_2 = 0$), and then to compute the response of the deforming structure from Equation (7). Cases where explosive waves act on structures are close to the "soft impact" response, while cases where the displacements u_1 and u_2 are of the same order of magnitude do not allow the uncoupling of the Equations (3) and (4), and are close to the "hard impact" definition. One may also classify these two limiting phenomena in a more simplistic manner. In the event of a "soft impact" of a deformable mass on a rigid structure, the kinetic energy of the impacting mass is transformed into plastic deformation of the impactor. However, in the case of a "hard impact" the impactor's kinetic energy is transformed into deformation energy in both the impactor and the structure. In this second case, if the impactor is assumed to be rigid and is arrested by the structure, its kinetic energy is transformed into deformation energy in the structure. Penetration will dramatically complicate these cases, and one must resort to numerical evaluations.

This discussion illustrates the degree of control that can be achieved by carefully selecting the corresponding parameters of the experimental set-up. Nevertheless, it is important to show what types of load pulses can be generated by impact testing, and their (qualitative) relationship with load pulses associated with HE detonations.

3 Examples of impact-induced loads

Impact testing can provide a wide range of load pulses which can be used in a very controlled manner. In order to illustrate this important feature, several examples of experimental load pulses obtained from various impact devices are presented and discussed.

Yan [8] studied steel-concrete bond under impact loads and used a 345 kg drop hammer with low drop heights (some as low as 0.3 m). These studies represent a lower end of impact testing capabilities, since the corresponding velocities and the localised (i.e. load applied to a single bar) nature of the load resulted in lower peak forces and longer pulse durations (Figure 1). Indeed, the loading rates that were achieved in those tests ranged from 5×10^{-5} MPa/s to 5×10^{-3} MPa/s.

Figure 2 contains three load-time histories obtained with the large pendulum at Penn State University (PSU) with a 2200 kg impactor. These data were obtained from test on structural concrete bridge parapets under vehicle crash conditions (i.e. the load pulses were designed to simulate the actual conditions by adjusting the shape, stiffness, and failure sequence of the impact bumper).

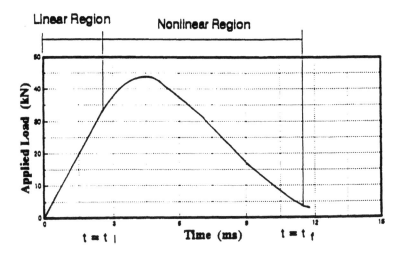

Figure 1. Low Level Impact Loads [8]

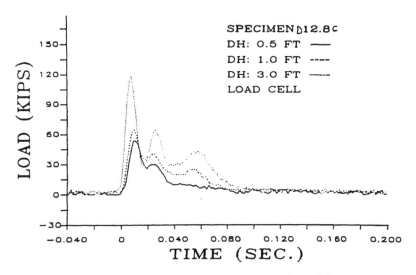

Figure 2. Load Pulses from the Large Pendulum Device at PSU

It should be noted that these data represent a very small range of loading pulses, since the drop height was only up to 0.91 m.

This pendulum can be raised to about 13 m, the mass can be varied between under 100 kg to 6,800 kg, and the bumper material and geometry can be varied from very soft to very hard. The use of special materials and shapes for the bumper will enable one to obtain a wide range of desired spatial pressure-time histories.

Thoma [6] describes several impact testing devices and methods that have been used successfully in Germany. These include both pendulum-type machines and drop hammers, whose general principles are similar to those proposed for this research.

Figure 3 includes two time histories from drop hammer impact tests (about 1100 kg at 13 m/s) on prototype and 1:8.5 model of a structural concrete T-beam.

It is important to note the good comparison between the load pulses for both prototype and model. The effect of the impactor's head shape on the pulse shape is illustrated in Figure 4 where flat and pointed drop hammer heads (about 1000 kg at 8.2 m/s) were used during tests on structural concrete slabs. Pointed heads result in a drastic reduction of the peak load, and a slight reduction in the duration.

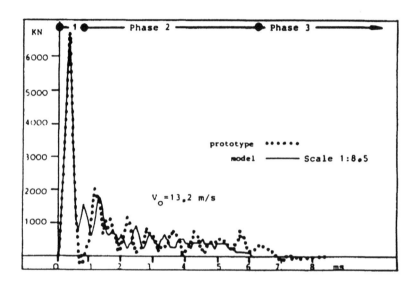

Figure 3.   Load Pulses from T-beam Tests [6]

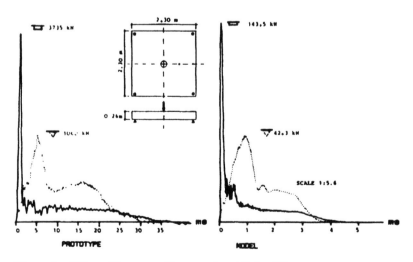

Figure 4.   Effect of Impactor Shape on Load Pulse [6]

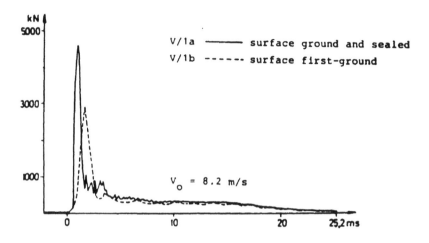

Figure 5. Influence of Surface Quality on Load Pulse [6]

Figure 5 [6] illustrates the effect of surface quality on the load pulse. It was found that an untreated concrete surface has a similar effect to that of a pointed impact head. Thoma [6] showed that the peak load is affected also by the concrete curing time (dryer concrete causes lower peak loads), concrete uniaxial strength (a 20% strength increase caused a 30% increase in the peak load), and by the size of the specimen (smaller sizes cause higher peak loads).

All these load pulses show a relationship between the impact conditions and the pulse shape. Low velocity impact will result in a smooth, long, almost half-sine wave, load pulse. As the impact velocity increases, the rise time shortens, the peak pulse increases, a main initial pulse with shorter duration emerges, and subsequent secondary pulses become less important.

Data collected during calibration tests for the drop hammer at PSU show the ability to obtain a wide range of load functions. The 26.75 kN (6000 lbs) hammer was dropped from different heights on a segment of steel rail attached to a pre-stressed concrete railway tie, and the impact interface and support conditions were varied. The load pulses were measured with a steel load cell which was attached to the impacting face of the hammer. Two accelerometers were mounted on the top surface of the railway tie: No. 1 at 250 mm and No. 2 at 1000 mm from the centre of the steel rail, respectively. Although several tests were performed, the data for two cases can be used to illustrate the potential of this test system.

The load pulse in Figure 6a was obtained for a drop height of 150 mm, the concrete tie was simply supported on 25 mm rubber pads and another 25 mm rubber pad was placed between the hammer and the steel rail.

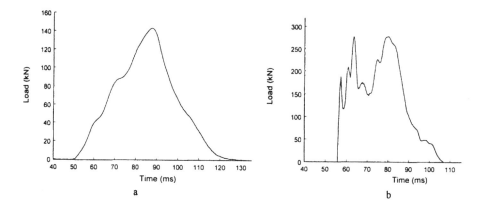

Figure 6.   Load Pulses from Drop Hammer Tests at PSU

The peak load of about 193 kN was reached at 38 ms after impact (3.76 kN/ms), and the pulse shape is quite smooth and triangular. The load described in Figure 6b was obtained for a drop height of 300 mm, the rubber pads were removed from the steel supports and from the interface between the hammer and the rail. Here, several local peak loads were observed. The first, 188 kN at 1.35 ms after impact (139 kN/ms), while the third was about 276 kN at 7.5ms after impact (36.8 kN/ms).

Under the first load pulse, the peak accelerations were about 13 g and 8 g for accelerometers 1 and 2, respectively (Figure 7a). The higher frequency signals that appear at about 120 ms were caused by the load cell slipping off the rubber pad and hitting the steel rail. Although this part of the signal is irrelevant to this discussion, owing to the accidental nature of its occurrence, it highlights the differences between different types of impact.

For the second load pulse, the peak accelerations were about 300 g and 200 g for accelerometers 1 and 2, respectively (Figure 7b).

In the frequency domain, for the first load pulse, both accelerometers exhibited power spectra in the range between zero to about 100 Hz, however, that for accelerometer 1 had a peak of about 0.13 at about 15 Hz while accelerometer 2 showed two peaks of about 0.038 at both 15 and 100 HZ (Figure 8a and 8b respectively). For the second load pulse, one could notice a significant difference between the two signals (Figure 8c and 8d respectively).

Accelerometer 1 had a power spectrum range between zero and 1700 Hz, with the major peak of about 0.38 at about 25 Hz, and several gradually lower peaks in the range of up to 500 Hz.

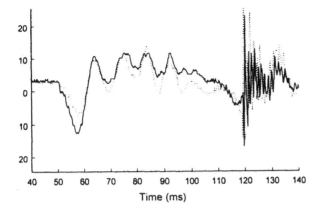

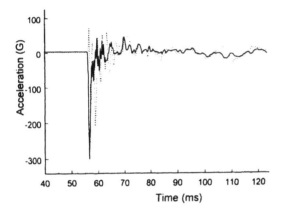

Figure 7. Acceleration Pulses from Drop Hammer Tests at PSU

Accelerometer 2 had a power spectrum between zero and about 650 Hz, it showed four distinct peaks in the range under 100 Hz (the highest of about 0.28 at about 75 Hz) and then three lower peaks at 120 Hz, 200 Hz and 570 Hz.

Clearly, one can study in great detail the characteristics of both the applied load and the structural response, thus deriving well defined relationships between cause and effect. Such data would be very valuable for the validation and verification of computer codes, since the analysts will be able to determine if, and how accurately, they were able to capture the same physical phenomena observed during the tests. Furthermore, from the physical data, numerical analysts would be able to assess the reasons for deviations between numerical and test data, and introduce modifications to remedy the problem.

It is important to note that this hammer can be dropped from a maximum height of 7 m, and the data shown below is at the very low end of anticipated loading pulses. Thus, one should expect a very broad range of physical data that will enable the validation and verification of computer codes over a wide range of both time and frequency domains.

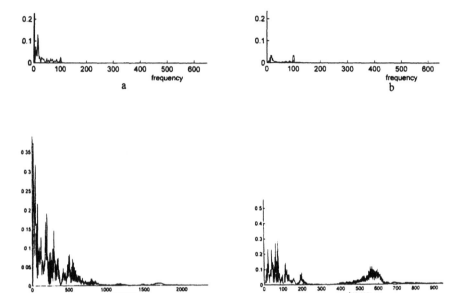

Figure 8.   Power Spectra for PSU Tests

For all the cases discussed above, the rise time varied from about 3.76 kN/ms to 23,000 kN/ms. Obviously, using higher velocity devices is expected to provide even steeper load pulses.   The selection of interface material (type and thickness) is expected to affect the load pulse, but this needs to be evaluated further.

As far as the negative loading phase is concerned, it has been shown that the effect of the negative loading phase from conventional explosives could be significant on lightweight structural elements (such as glazing and cladding). However, based on the discussions during the NDCS-sponsored workshop in 1993, its precise role is not well understood, and it is unlikely to be very significant when heavy structural concrete systems are considered under the same type of environments.   Therefore, it is expected that the proposed precision tests should provide meaningful data in support of code validation and verification.

4 Reproducibility of test results

It is important to address the issue of reproducibility, since it is one of the important reasons for initiating this research effort. Supporting information is found in the references cited above. Obviously, assuming a well maintained testing facility, there is excellent control on the potential energy levels which propel the impact device toward test articles. Nevertheless, careful attention must be given to specimen preparation in order to ensure comparable load pulses, as shown by Thoma [6], and briefly mentioned above. Under such conditions there will be little variation (under 10%) on the peak load values. Based on the supporting information from other impact testing facilities, the reproducibility of such tests is assessed as very good, and repeated testing (at least three tests per specimen to ensure load pulse stability) and cross correlation between testing organisations will ensure data objectivity.

5 Structural testing and numerical analysis facilities at PSU

In addition to dynamic materials testing facilities at PSU, structural dynamics testing is performed at the Structural Engineering Laboratory at the Department of Civil and Environmental Engineering. Truck access is provided for the laboratory, and a crane system is available for moving equipment and models to any point in the laboratory. The main test bed of the laboratory consists of load points, flush with the floor slab, designed to carry loads of 445 kN each and spaced at 1.52 m centres. It can accommodate models up to 16.75 m in length and 4.25 m in height. Walls adjacent to the test bed are designed to resist lateral loads of 44.5 kN, and wall load points are spaced at 1.52 m centres. A large adjustable loading frame is used with the test bed which permits beams and model frameworks to be loaded at any location on the test bed, and at any position on the test specimen of model. Structural steel stiffbacks also permit lateral loads or lateral support.

A 1.33 MN universal testing machine is located in the laboratory. Sufficient tools and equipment are available for fabricating welded or bolted structural steel models, and an adjacent laboratory is well equipped for casting and curing concrete specimens and for performing all types of concrete testing. The laboratory is equipped with Riehle-Los fatigue testing equipment capable of applying repeated loads at speeds between 100 and 1000 cycles per minute. The corresponding loads could be up to 1.42 MN.

Structural testing under short duration dynamic loads can be performed on several devices at, or associated with, the laboratory. The large-scale outdoor impact pendulum facility consists of a 15.25 m high steel frame capable of swinging weights of up to 67 kN through an arc with a vertical drop height of up to 13 m. Additional advanced impact testing devices have been recently added with support from the National Science Foundation and the University: An indoor impact pendulum facility consisting of a 4.25 m high steel frame capable of swinging weights of up to 7.1 kN through an arc with a vertical drop height of about 3.65 m, and a 8.9 m 26.75 kN drop hammer device with a 7 m drop height. These devices are expected to be modified over time to further enhance their capabilities. Several multi-channel high speed (up to 1 mega samples per second) data acquisition systems support these testing devices.

Additional high precision testing devices for short duration dynamics include an Instron 1331 dynamic testing machine. The actuator can deliver a maximum force of 19.62 kN with a maximum stroke of +/- 165 mm. The dynamic testing rate for the closed loop system is up to 17 m/s (Actually, a velocity of over 23 m/s was obtained, and the device can be operated also in an open loop mode at faster loading rates). As far as data acquisition is concerned, these systems are supported by various very high speed data collection systems (up to 1 MHZ) for strain, displacement, force etc., including a 12,000 frames per second SP2000 camera and a copper vapour laser system (a one million frame per second camera in the College of Engineering can be used also if required).

Several advanced computer systems exist at PSU. At the Centre for Academic Computing (CAC) various clusters of engineering workstations (based primarily on Sun, Silicon Graphics and IBM RS6000 systems) are currently used for most advanced computational efforts, and a 48 processor IBM SP2 have significantly enhanced these capabilities. These facilities are expected to be enhanced further in the near future. An IBM 3090S at CAC and a Cray YMP at the College of Earth and Mineral Science can be used in support of such activities. Links to several supercomputer facilities (such as at Cornell University and at the University of Pittsburgh) are available, and are used for massive computations.

Although a large number of finite element codes is available at CAC, the numerical efforts in the short duration structural dynamics area has focused on the following codes: DYNA3D, ABACUS, EPIC-95, CTH and a family of special codes developed by Professor Krauthammer and co-workers. These special codes include a sophisticated SDOF code for preliminary non-linear structural behaviour evaluation (this code has been distributed to various DOD organisations), a unique finite difference code (based on the Timoshenko beam and Mindlin plate theories) for careful examination of behaviour and failure, a hybrid finite element – finite difference code for medium-structure analyses (developed under a DNA sponsored study), and a combined symbolic-numeric code for post-event damage evaluation. Several new constitutive models have been developed by Professor Krauthammer and co-workers, and they have been incorporated into these codes. Internet access to other sites (such as the Cornell Theory Center) have been used for various applications.

6 Conclusion

In today's environment (which will probably continue into the foreseeable future) the need to protect the civilian population against social unrest seems to far exceed the previous role of military-sponsored work on fortifications. Unlike the global politically/ideologically motivated conflicts of the past, most of the armed conflicts in the last few years have been dominated by social, religious and/or ethnic localised conflicts. It may no longer be the traditional "them against us" type of situation, but a new and evolving "them" has to be considered. Therefore, the future of R&D in the area of short duration dynamics has to be shaped accordingly. The need for large-scale testing of heavily fortified military facilities may no longer be the main area of concern (although the technology in this area must be kept relevant) and careful attention must be devoted to typical civilian facilities whose failure could severely disrupt the social and economic infrastructure of nations. There is a serious lack of knowledge on how such facilities (office buildings, schools, hospitals, power stations etc.) will behave under blast and shock loads. Many of the materials and components were never studied for such applications, and most nations do not have the required resources to approach this problem in the "old fashioned way".

Future R&D in this area relies heavily on the development of precision testing methods which are aimed at supporting computer code validation and verification. Furthermore, a strong multi-national effort is urgently needed to address this challenge, and to ensure that sponsoring organisations will be both guided and educated to move in this direction. The collaborative R&D activities will insure that the parallel work will be well coordinated, and that the resources will be efficiently used. Cross-checking and mutual assessment of methods and data will ensure that the maximum amount of knowledge will be extracted from such data, and be wisely implemented to protect the public against serious consequences.

7 References

1. Krauthammer, T., June 1993, *Structural Concrete Slabs Under Impulsive Loads,* Fortifikatorisk Notat Nr 211/93, Norwegian Defence Construction Service.
2. Eibl, J., September 1987, *Soft and Hard Impact,* Proc. 1st Int. Conf. on Concrete for Hazard Protection, Edinburgh, Scotland, pp. 175-186.
3. Feyerabend, M., 1988, *Der harte Querstoss auf Stützen aus Stahl und Stahlbeton,* Institute für Massivbau und Baustofftechnologie, Karlsruhe, West Germany.
4. Krauthammer, T., December 1989, *Reactive Protection for Hardened Facilities,* Final Report No. ESL-TR-89-34, Air Force Engineering and Services Centre, Tyndall AFB, Florida.
5. Krauthammer, T., November 1994, *Buildings Subjected to Short Duration Dynamic Effects,* Proc. First Cardington Conference, Building Research Establishment, Bedford, England.
6. Thoma, W.W., 1992, *Modelling Structures Subjected to Impact,* Ch. 10 in Small Scale Modelling of Concrete Structures, edited by F.A. Noor and L.F. Boswell, Elsevier Science Publishers, pp. 273-309.
7. Bischoff, P.H., June 1993, *Analysis of Concrete Structures Subjected to Impact and Impulsive Loading,* Proc. Annual Conference, Canadian Society of Civil Engineering, pp. 255-264.
8. Yan, C., October 1992, *Bond Between Reinforcing Bars and Concrete Under Impact Loading,* Department of Civil Engineering, University of British Columbia, Vancouver, Canada.

8 Acknowledgement

The studies supporting this paper were partially supported by NATO Collaborative Research Grant No CRG 950283.

X-RAY SCREENING OF MAIL AND DELIVERED ITEMS & THE CARDINGTON LARGE BUILDING TEST FACILITY

C.J.R. VEALE
Government Security Advisor, Ministry of Defence, UK

1 Background

The aim of this paper is to present proposals for a series of explosive trials utilising the LBTF for the development of a Mail Screening Facility. The facility, which is to be constructed using lightweight materials as would generally be required if it were to be built into an existing modern office building, is designed to minimise the consequences of blast and fragmentation effects on the operator and post room staff from an explosive device detonating within the X-ray machine.

Over the past few years there has been a measurable increase in the number of incendiary devices, explosive devices and dangerous articles that have been sent through the postal/delivery systems by terrorist, political and criminal groups.

This increase in the number of devices sent, together with an apparent improvement in the sender's ability to disguise the devices (e.g. through the use of video cassette boxes, postal tubes, jiffy bags etc.) has made detection by simple visual examination very much less reliable.

2 The threat

During the 1980's and early 1990's the number of such recorded devices sent each year was fairly evenly divided between terrorist groups (predominantly the Provisional IRA) and Animal Rights activists. When PIRA called a "cease-fire" on the 1st September 1994, their use of the 'letter bomb' ceased. The remaining groups, however, particularly animal rights activists, continued to use them in increasing numbers, maintaining the general upward trend.

Other groups who used letter bombs were:

- The Scottish National Liberation Army (SNLA)
- Criminals
- People with grudges

It is extremely difficult to assess accurately the future threat. Likely areas/groups from which a potential threat may occur are:

2.1 Domestic

- Irish Terrorism (should the peace process fail)
- Animal Rights Extremists
- SNLA
- Criminals
- People with grudges
- Those of unsound mind
- Copycat activity

2.2 International

- Middle East – Islamic Extremists
- Iraq
- Balkans
- Algeria

3 Responsibility for detection

It is the responsibility of the recipient to undertake the detection of dangerous items and devices delivered to him. It is not the responsibility of the Post Office, courier or delivery company, although they may intercept a suspect item if they recognise any visual external characteristics. However, none undertakes routine X-ray screening.

Therefore, those who are under threat, or who consider they might be in the future, should create a post-screening facility manned by properly trained and supervised staff.

4 The trials

The cost of dealing with this perceived risk clearly could add to the overheads of the company or institution involved. This financial consideration, together with the need to protect staff from injury, and business from interruption, requires that investment in security systems should achieve the required protection at an acceptable price.

To assist in the development of advice for architects and others in the construction industry involved in the design and installation of such facilities, two similar postrooms based on one of the government's standard layouts for a continuous screening system are to be constructed at the BRE's LBTF. One room will be constructed using a standard manufactured high-duty partition walling system (e.g. British Gypsum's Glasroc blast refuge wall – Figure 1) erected in accordance with the manufacturer's specification. The other will be constructed using the same partition materials, but to an enhanced specification.

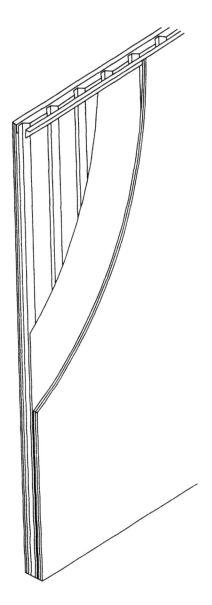

Figure 1.   A schematic of British Gypsum's Glasroc blast refuge wall

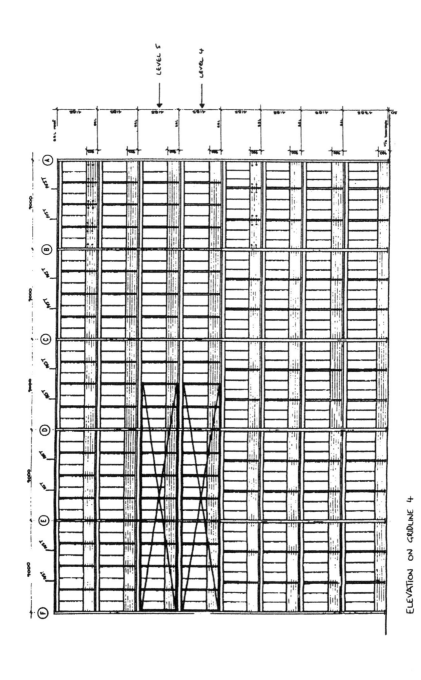

Figure 2. Rear elevation of the building showing the locations of the two postroom facilities

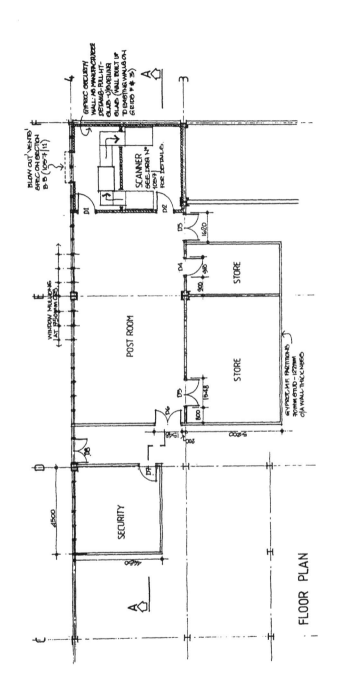

Figure 3.   Plan of the postroom facilities

The location of these two sets of rooms is on the fourth and fifth floors of the eight-storey building, adjacent to the rear wall (Figure 2). Of course, post rooms would not normally be located in such positions, but since the building is founded on a 1.25 m thick concrete foundation and usually such buildings would have basements, and hence flexible floors at ground level, it was felt that simulation of this condition was important for the tests. Another criterion to be considered was the surrounding structure for the two facilities. Here the need was to ensure similarity, and, because of the local damage caused in other programmes, this is best met by choosing the higher locations.

The design and location of the room sets have been undertaken by an architect with the brief that the building is required to house the headquarters of a national law firm with a staff of around 300. Suspended ceilings with light fittings and hidden forced ventilation ducts will be installed, as will raised floors which are typical in this scenario.

The external gable walls are 140 mm dense concrete blocks lined with heavy duty partition panels, whilst the main flank walls are glazed with a proprietary system installed over a 600 mm high concrete block dado wall. The scanner room section will be designed with a glazed vent in at least one of the tests.

The intention is to create the best replication of a real life situation – but without the live bodies!

The principal tests will involve the detonation inside an X-ray machine casing placed in the Scanner Room (Figure 3). Of particular importance will be the opportunity to observe the performance of the walling systems, especially their capability to contain high velocity fragments created by the device itself and the X-ray screening machine. There are other characteristics of this hazard which will be studied in the programme, notably the accuracy of the predictive tools for pressure distribution. Often the computer codes for dynamic load assessment have not been calibrated against data from such complex, but realistic, spaces, and the high quality data from these tests will be invaluable for this purpose. Many explosion tests have been undertaken in simpler environments, but these tests will be unique in their variety and relevance to real life.

It is envisaged that full instrumentation, recording displacements and time/pressure histories, will be undertaken. There will also be video and still records, and, of course, measurements of the dynamic response of elements and the structure.

The output from this programme will be improved advice for designers and validation of some hazard assessment methodologies.

PART FOUR

Other Buildings

BUSINESS PROCESS ANALYSIS AND ITS POTENTIAL APPLICATION TO CONSTRUCTION

J.H. ROGERSON and P.J. DEASLEY
School of Industrial and Manufacturing Science, Cranfield University, UK

1 Introduction

The importance of the UK construction industry is beyond doubt. It employs over one million people, contributes 9% of GDP and in 1993 realised contracts worth some 3.2 billion pounds. It is number four in Europe (behind Germany, France and Italy) and represents over 10 per cent of European Union construction activity.

However, the recession at the beginning of the nineties has had a severe impact on the industry. It is falling behind European and worldwide competitors and yet is felt to have considerable potential for growth. In comparison with other industries, for example manufacturing, improvements to competitiveness have been poor.

A series of major reports has studied the construction industry. The Latham Report [1] concluded that construction costs could be reduced by 30% if concepts from other industrial sectors were introduced (concurrent engineering, just-in-time supply, customer-supplier partnerships, Total Quality Management, etc.). The Miller Report [2] introduced "Innovative Manufacturing" based on a business process approach.

The concept of 'Construction as a Manufacturing Process' was discussed and has subsequently become a major thrust in the industry-led EPSRC research programme, The Innovative Manufacturing Initiative (IMI). Technology Foresight, reporting the results of its construction panel [3], has also highlighted the need for a business process approach to radically re-engineer the construction process.

Over the past decade there has been a significant renaissance in the performance of the UK's best manufacturing companies, some of which have achieved world best practice. This performance is usually measured in terms of the cost, quality and delivery of products and services.

The opinion is growing that the time is ripe for radical improvements to these same parameters in construction. A business process approach is thought to offer the greatest potential for success.

2 Modern manufacturing

The corporate strategy of major manufacturing companies has swung away from ideas of market position in relation to competitors (Porter's Competitive Strategy, [4]) towards a focus on the customer (Porters Value Chain Theory, [5]) Serving the customer by efficient delivery of a quality product giving value for money demands efficient business processes within the manufacturing organisation. In general the two dominant product delivery processes are current product to customer and new product to market. However, these key processes will depend on a raft of support processes to achieve acceptable performance.

Modern manufacturing, sometimes referred to as 'lean production', (in comparison with mass production of the Henry Ford era), is characterised by flexibility to produce customised variants in small batches (even batch sizes of one!).

The business achieves world class performance by being process focused (rather than in functional groups). This involves multi-disciplinary teams, concurrent (not sequential) working, flat organisation structures and bench marking performance against competitors and other industries.

The business is, in effect, an extended enterprise where the manufacturing company, its customers and its suppliers, are woven together by information sharing.

The customer-supplier relationship (usually with 'partner' supplier companies) is crucial to the organisation and central to the just-in-time supply system which has allowed inventories to be drastically reduced and capital released for business improvement. In essence it is supply chains that now compete rather than individual companies.

3 The business process approach

The concepts of Business Process Redesign originated in a study at MIT from 1984–1989 entitled 'Management in the Nineties' [6]. This stressed the importance of information technology (IT) in the efficient management of modern organisations. From this developed a number of related concepts, for example, Business Process Re-engineering highlighted in the seminal paper by Hammer [7]. Business Process Redesign usually refers to the internal workings of a single organisation. In modern manufacturing, Business Network Redesign is often of greater meaning as it refers to supply chains [8].

4 The role of information technology

The complexity of a business process approach in an extended enterprise is rendered workable by the employment of appropriate information technology. IT is the integrator between people, organisations, operating sites and functional departments. It can take many forms, from communication systems (e.g. Electronic Data Interchange, EDI) to data management systems (e.g. Engineering Data Management, EDM) to functional linkages (e.g. CADCAM) and design aids (e.g. Design for Manufacture and Assembly tools, DFMA). The application of IT is leading towards a "virtual environment" where companies, prototypes and processes can exist in electronic rather than material form. This paperless virtual enterprise poses a great management challenge for the next decade. It promises rapid, efficient product delivery processes for future customers.

5 Benefits realised by a business process approach in manufacturing

Business Process Re-engineering has been grasped by manufacturing and has produced some dramatic results. A flavour of the benefits on offer can be obtained from the following examples [8]:

Pilkington Optronics: a manufacturer of advanced optical components and systems.

- Manufacturing lead-time reduction   50%
- Stock and work in progress reduction   70%
- Sales per employee increase   250%

Lucas Automotive: a major supplier of automotive systems and components.

- Manufacturing lead-time reduction   80%
- Operating cost reduction   25%
- Quality level improvement   factor 10

It must be remembered that behind these dramatic improvements lie fundamental radical changes to well established operating businesses. The implementation of change is never easy, and change to achieve step-function improvements is usually traumatic and a question of survival.

6 Application of business process analysis to construction

It would be naive to think that a radical change in enterprise-management principles can be easily transported between industry sectors. Automotive industry is fundamentally and culturally different from construction, and it is not obvious that changes in the former are relevant to the latter. Although a business process approach has been suggested for construction, it has yet to be demonstrated.

Manufacturing industry is characterised by a stable environment; an established factory, key sub-contractors and component suppliers knitted together in established supply-chains, a stable product range and, often, known customers. In contrast a construction project is usually unique in terms of product, site, client and sub-contractors. However, construction is an assembly business like automotive and aerospace. Its business processes are no more complex than manufacturing but tend to be combined and managed on a project to project basis. It is therefore more difficult to identify best practice and benchmark activities between companies.

The procedures for the identification of construction processes is not different from that used elsewhere, and the same mapping tools (e.g. input/output analysis,

IDEFO, etc.) should be useable. As with manufacturing there is an inherent danger of defining a process too tightly to allow meaningful improvements to be made. The key feature is to focus on the process and to avoid focusing on the sometimes complex functional groupings or the contractor/sub-contractor interfaces and responsibilities.

Examples of delivery processes in construction are the erection of a concrete frame and the installation of M and E Equipment. These might be supported by processes covering cost report preparation and establishment and maintenance of site services.

Delivery processes have a specific desired outcome as the end point i.e. a part of the building. Support processes enable a delivery process to be carried out. By analysing the processes in this fashion, non value-adding activities can be identified and eliminated. We can then consider process simplification and perhaps integration, and, where appropriate, we can apply IT to underpin the improved process. It must be pointed out that the labeling of an activity as value-adding or non value-adding is not clear cut and often leads to significant reassignment of roles and redefinition of activities across the whole supply chain. The final requirement of a business process approach is the definition of suitable performance measures such that continuous improvement can be quantified.

7 Some potential problem areas

The problems arise from the nature of the industry. Business process analysis is time consuming and requires skilled resources. The uniqueness of the product and its environment mean that industry must undertake a form of process analysis for each project.

The necessity to make fundamental changes to established contractual attitudes or even contractual responsibilities, to ensure the working of the extended construction enterprise is, furthermore, no trivial matter. There also exist a host of problems that spin-over from manufacturing. These include team recognition and rewards, accountability in a less hierarchical organisation and focus on a complex client.

The one-off nature of the construction business underlines the need to seek out generic forms of business process re-engineering to avoid resource and time consuming analysis on each new construction process.

8 Potential benefits

These have yet to be demonstrated but, by analogy with manufacturing industry, they could be expected to be in the region of:

- project cost reduction   40%
- quality improvement   factor 10
- delivery improvement   50%

There are also several significant secondary benefits which include:

- establishment of a world competitive industry
- generation of a trained and motivated workforce
- drastically reduced wastage
- better value for clients
- fostering of an atmosphere of mutual trust
- development of a system of shared and accurate information

9 Conclusion

Construction is a vital industry sector for the UK and we must reverse the recent decline in market performance. Business process analysis and re-engineering offer the potential to achieve the significant step-function gains already achieved in manufacturing. Today, some of our manufacturing companies are world-class; tomorrow some of our construction companies must also be world-class in order to survive in a global market place. In considering a business process approach we must decide:

- to do it
- to do it now
- to continuously improve

We can then guarantee the survival of a vital section of UK industry.

10 References

1. *Constructing the Team*. HMSO 1994.
2. *Innovative Manufacturing Initiative: A New Way of Working*, SERC 1994.
3. Technology Foresight: Progress through Partnership. Report for Construction Sector, HMSO 1995.
4. Porter. M.E., *Competitive Strategy*, The Free Press 1980.
5. Porter. M.E., *Competitive Advantage*, The Free Press 1985.
6. Scott Morton M. (Ed.), *The Corporation of the Nineties: Information Technology and Organisational Transformation*. Oxford University Press, New York 1991.
7. Hammer. M., *Re-engineering Work: Don't Automate – Obliterate*, Harvard Business Review, July-Aug 1990.
8. Peppard J., Rowland P., *The Essence of Business Process Re-engineering*, Prentice Hsil 1995

PERFORMANCE RELATED RESEARCH ON THE IN-SITU CONCRETE BUILDING

P.S. CHANA
Director of CRIC, Imperial College, UK
S. ARORA and R.M. MOSS
Concrete Structures Division, Building Research Establishment, UK

1 Introduction

The initial focus of the European Concrete Building Programme (ECBP) is geared towards the construction in 1996 of an in-situ concrete building inside the Cardington Laboratory. This phase will incorporate a number of innovative features in the building and its construction, and process related research, both on and off site, described in previous papers. The aim of this work will be to re-engineer the whole process of design and construction of in-situ concrete frames as an 'integrated business process'. In line with the Latham recommendations and the IMI CMP Sector Target, the objective is to increase client value, and so the Concrete Industry's competitiveness, by 30% through lower costs, improved speed and quality. A two year programme and comprising several projects, this phase will deliver a restructured process, best practice guides for construction of in-situ concrete frames and critical IT interfaces.

The design of the frame which is in accordance with EC2 is also a demonstration of buildability for fast and efficient concrete construction. The following features are included:

- efficient floor construction
- unobstructed soffits
- avoidance of edge upstands
- use of high strength concrete to standardise column sizes
- elimination of concrete walls
- use of eccentric steel cross bracing for stability
- zoning of service hole requirements
- rationalised reinforcement and prefabricated mats

Once the process research has been completed, this building offers a unique opportunity for testing the behaviour of a concrete frame under service and ultimate loads taking into account continuity effects and whole building behaviour.

The question may be posed: Why should we test such large structures? Armer and Moore [1] have argued the case for full scale testing. A large amount of research into the behaviour of structures has been concerned with the structural performance of isolated members, subassemblies, and the development of analytical techniques. Local failure in structures is generally studied by large-scale component testing, while the overall behaviour of the structural system is investigated using scaled-down experimental models. Such tests are designed to provide basic data for model development and verification. Simplified test structures cannot, however, truly represent the behaviour of a complicated building fabricated and erected under normal commercial conditions, carrying floors, walls and encased with cladding. Experience has shown that the assumptions related to boundary conditions for isolated element tests are often so unsatisfactory that the apparent properties of the elements are only poorly deduced.

So many differences from the behaviour of isolated component parts arise when they are connected together that questions concerning the force redistribution capability of highly redundant structural systems cannot be answered by component testing. Furthermore the global and local failure behaviour of the building and the effectiveness of both structural and non-structural repairs can be proved only with tests on a number of different types of completed building.

This paper is a summary of performance related studies which may be carried out on the in-situ building. It is the basis for identifying a number of firm projects to be developed into full research submissions for funding through EPSRC and other sources. This research should lead to an improvement in our understanding of the performance of full scale concrete buildings and the development of analytical and design methods for advanced design as well as current codes such as EC2. Each project has specific requirements for floor space, instrumentation, applied loading and experimental data to be collected since many of the performance projects are interrelated with overlapping requirements and a major co-ordination exercise is planned over the next few months.

2 Description of building

2.1 Structural form and basis of design
The in-situ concrete building consists of a 7-storey flat slab frame with a 3 by 4 bay layout based on a column grid of 7.5 m by 7.5 m. The slab thickness is 250 mm. Provision is to be made for normal building services including air conditioning to modern office standards. Characteristic imposed loads are as follows:

Office Areas;	Imposed	2.5 kN/m^2
	Partitions	1.0 kN/m^2
	Total	3.5 kN/m^2
Roof;	Imposed	1.5 kN/m^2
Plant Areas;	Imposed	7.5 kN/m^2

Dead loads in addition to the self weight of the structure are as follows:

Office areas;	Ceiling/Raised Floors & Services	$1.0 \, kN/m^2$
Roof;	Finishes, Ceiling & Services	$3.0 \, kN/m^2$

The structure is designed to Eurocode 2, supplemented where necessary by BS 8110 and other British Standards. The structure is designed as a braced frame; bracing is in the form of steel X or K members to be installed asymmetrically in the structure to facilitate research on lateral loading and whole building behaviour. The design allows for main elevations to be finished with precast concrete cladding with insulation and dry lining. Total load allowed for is 8.85 kN per linear metre on each floor.

The structure is founded on 1500 mm square pads on flexible raft sub-base with an allowable bearing pressure of 100 kN/m^2.

In-situ concrete to floor slabs is to be grade C30/37 (cylinder strength 30 N/mm^2 cube strength 37 N/mm^2) as defined in Eurocode 2. In practice, a C40 grade will be specified.

Floors 1 to 3 will be reinforced conventionally with a standard C40 grade mix. Floors 4 to 6 will be a modified concrete mix of grade C40 to investigate fast drying aspects for one of the production project. The intention at the moment is to provide a higher strength concrete (Grade C70/85) for Floor 7 which is the roof slab.

Columns at lower three levels will use Grade C70/85 changing to C30/37 to maintain constant column sizes.

2.2 Allocation of floor space

In a project of this nature, it is vital to co-ordinate the requirements for floor space required for the various research projects. Details of this are to be finalised after discussion with the project leaders but some general decisions have to be taken in this regard at this stage.

Floor 1 will be used for large scale fire testing on the ground floor after the performance tests have been completed. Floors 2 and 3 have built-in instrumentation including strain gauges and VWG's and will be used for static load testing under serviceability and ultimate loads. Floors 4 to 7 can be used for other performance testing projects if funding permits but no built-in instrumentation has been provided.

2.3 Built-in instrumentation

2.3.1 Measurement of reinforcement strain

The instrumentation to be provided in the inset concrete building is being finalised and preliminary drawings have been produced indicating where strain gauges are required to be installed during construction on the top and bottom reinforcement in the slabs on the 2nd and 3rd floors.

The two basic types of strain measuring device which can be used are electrical resistance strain gauges (ERS gauges) and vibrating wire strain gauges (VW gauges). It is proposed to use one of each type of gauge at selected locations to measure the reinforcement strain.

The use of ERS gauges is well-known. The type of VW gauge which would be used to measure reinforcement strain would be a surface mounted type. This is screwed to mounting plates welded to the surface of the reinforcement with the standard gauge length of 5.5 inches. The strain in the reinforcement and the adjacent concrete should be the same; nevertheless, to try and ensure that the steel strain is measured and not the concrete, the gauge can be isolated from the surrounding concrete for example by wrapping the gauge and the bar to which it is attached with a flexible material such as rubber.

The VW gauges come pre-assembled with any required cable lengths. For both types of gauges the cables can be protected by running them through electrical conduit. It is particularly important to provide this protection where the cables emerge through the surface of the concrete.

Data logging systems are available which can deal with measurements taken from both the VW and ERS gauges. The standard logger typically has a capacity of 30 channels but this can be extended using expansion modules to give 60 to 90 channels if required.

A possible alternative method is to put load cells in series with particular reinforcement bars. These would need to be sleeved to eliminate concrete bond and transfer of shear stresses along the side of the load cell, but would give a direct measurement of the force (and hence stress) in any particular bar. However, whichever approach is used, there will always be some localised effects because of the disruption of the bond to the concrete.

If funding permits, it is hoped to incorporate some special reinforcing bars developed at Durham University with electric resistance gauges in a duct running longitudinally though the centre of reinforcement. These will be incorporated on floors 2 & 3 at critical column slab junctions.

2.3.2 Measurement of concrete strains

Measurement of concrete strains using embedment type VW strain gauges is more established. The comments relating to cabling and logging of the surface mounting type gauges apply equally to the embedment gauges. The location and number of gauges to be provided shall be determined by the research requirements and funding available.

2.3.3 Measurement of temperature

Thermocouples for temperature measurements are to be provided on all floors, and can be made from standard thermocouple wire (e.g. copper/constantin). The wires can be fed directly into a data logging system which will output the measured temperature based on the characteristic of the particular wire used. The data logging system described above can also handle thermocouples. There appears to be no particular problem with cable lengths with thermocouples. Hence it may be possible to combine the temperature measurements with those from the vibrating wire strain gauges in a single system.

If it is only required to measure the temperatures reached during the hydration process, the exact positioning of the thermocouples within the depth of the slab may not be critical. Nevertheless protection will still need to be provided to the wires where they emerge from the concrete. Past experience suggests that 19 s.w.g. thermocouple extension cable would prove a suitably robust type of cable to use.

2.3.4 Instrumentation required for fire research

The instrumentation required for this work will be installed subsequent to construction. Additional thermocouples will be installed by backfilling cores drilled through the cast concrete. However void formers are required to be included when casting the concrete which can be subsequently be removed and strain gauges mounted on the reinforcing bars. FRS is developing details of the location of void formers and instrumentation.

2.4 Loading sequence
The performance projects require static and dynamic loads to be applied. A possible sequence of static loading is given in Table 1 below. Where necessary, dynamic loads can be applied in conjunction with the static loads. In general, fire and explosion testing will be carried out after the static loading tests have been completed.

Table 1. Example of static loading sequence

Load Case	Loading	Load (kN/m^2)
1	Self weight	6
2	Service load	7.5 (1.5)
3	Characteristic Imposed Load	9.5 (3.5)
4	2 x Characteristic Imposed Load	13 (7)
5	To shear failure (selected locations only)	

Figures in brackets are the additional loads to be imposed.

Load cases 1–4 can be provided by means of sand bags (up to 11 kN/m^2 applied loading). Load case 5 can be provided through an arrangement of McAlloy bars around the column.
   Detailed analysis is to be carried out to assess the flexural and punching shear strength of the slab prior to finalising the loading sequence.

3 Serviceability performance

3.1 Design of thinner flat slabs
For serviceability reasons, deflections in flat slabs are restricted to span/250. Generally, half of this deflection is considered to be short term and the remainder long term due to creep. In BS 8110, slab deflection is controlled by simple rules that restrict the span/depth ratio to typically between 25 and 30 for continuous slabs. This is an empirical rule which has largely remained unchanged in the UK since reinforced concrete construction began. Concrete strengths have increased with a corresponding

rise in stiffness characteristics and the beneficial effects of this are not considered in current UK design practice.

The whole area of slab deflections needs to be reviewed by considering international design practice including EC2, surveying span to depth ratios in existing structures and developing an accurate computer based method of calculation. The design method should take into account the following factors:

- layout of slabs and supporting system
- level of applied loading
- actual properties of concrete including stiffness, tensile strength, creep and tension stiffening

It is considered that slab depths could be reduced by as much as 25% in many situations which would extend the allowable range of span to depth ratios to between 30 and 35.

The proposed in-situ reinforced concrete framed building at Cardington, provides ideal conditions for measuring the deflections of flat slabs during construction and under known service loads over a period of 2 years. This information would be used to verify and further develop the proposed design method. Deflections can be measured on various floors with different reinforcement arrangements using 3D laser survey systems and accurate levelling for a separate check. This will give accurate data on the performance of full scale slabs under different service stress. It is planned to construct one of the floors in high strength concrete which will enable the analytical models to be calibrated for these concretes. Control specimens should be taken to monitor the properties of concrete under sustained loading.

3.2 Dynamic characteristics of thin flat floors

As slab depths are reduced, it is vital to check the dynamic characteristics under normal and overload situations. It is planned to carry out induced vibration tests aimed at determining the dynamic characteristics of thin/slender flat slab floors with varying boundary conditions provided by the building. The data on the natural frequency and mass will be provided which will be extremely useful in the serviceability design of such floors, both under normal loads, e.g. people walking, car park loading etc, and accidental dynamic overload situations, e.g. dance floors.

3.3 Dynamic performance of structures

Tests are planned to measure the overall dynamic characteristics of the flat slab building structure with no shear provision. The tests will introduce forced vibration of the building induced by large vibrator systems installed at the top and on different levels of the structure. Data will be obtained on the natural frequency of the modes of vibration, mode shapes, coupling of modes, overall stiffness and damping characteristics. The information thus provided will be used to verify theoretical models that predict the overall behaviour of the building. These tests will also be repeated when cladding, finishes etc are installed.

The natural frequency is the most important item where the seismic force for a structure is to be predicted. This data will also provide valuable information on earthquake design.

The vibration characteristics of the building will be studied during construction as the height is increased. These tests will not need forced vibration but will measure the fundamental frequencies of the structure using remote Laser interferometer. Opportunity will also be taken to carry out further tests aimed at determining the dynamic characteristics of the building as follows:

- investigating the change in the overall response of the building with localised structural damage. Here the vibration monitoring will be undertaken after local structural damage has occurred, e.g. local static overload, after fire tests and after gas explosion testing.
- investigating the response to ground borne vibrations. The work that is planned is to provide a source of vibration between the source and positions on the building and the ground. The source conditions that are envisaged are first, continuous vibration from a vibration generation and second, impact vibrations from some disturbance on the ground.

3.4 In-service Monitoring and Smart Structures Technology
The in-situ building at Cardington could provide an opportunity for testing out novel instrumentation techniques, e.g. use of optical fibre sensors for strain measurement.

This would be in addition to the instrumentation already cast into the structure and other instrumentation to be installed subsequently, i.e. vibrating wire gauges, electrical resistance strain gauges and thermocouples.

The instrumentation will provide a dual role – giving required measurements for specific tests for which it has been installed but also providing general data for monitoring purposes for example during the jacking tests required for the study of differential settlements.

3.5 Reinforcement strain and bond stress measurement
A detailed knowledge of reinforcement strain distributions is a fundamental prerequisite to understanding the behaviour of reinforced concrete structures, particularly the intrinsic mechanism of load transfer by bond. This has prompted a number of attempts to obtain this data using a variety of experimental procedures, the most effective being to install electric resistance strain gauges in a duct running longitudinally through the centre of the reinforcement. This avoids degrading the bond characteristic of the bars (by leaving the surface of the bars undisturbed) and permits very detailed measurement to be made. The technique has been considerably developed and improved at Durham University over the last twelve years to become a powerful and flexible tool for use in reinforced concrete research.

Some reinforcing bars of this type should be incorporated in the column slab connection regions on Floors 2 and 3 to enable the development of strain in the reinforcement and hence the bond stress transferred to the concrete to be studied. This data will form useful input into the other performance related projects such as punching shear and redistribution of moments.

4 Analysis and design

4.1 Loadings on concrete floors

The British Council of Offices (BCO) has recommended that the imposed load for offices of 2.5 kN/m^2 as specified by BS 6399 is appropriate but many developers still prefer to use 4.0 kN/m^2 which became normal for office construction in the 1980's. Load testing under service and ultimate load conditioning will be carried out on the structure to demonstrate the reserve of strength in flat slab floors. This testing should also demonstrate adequate performance under the larger service loads which can occur in areas such as corridors. This project in closely linked with the other studies planned on analysis, design and performance.

4.2 Automated yield line analysis of floor slabs

Automated yield-line analysis is a technique for investigating the ultimate load carrying capacity of a slab in flexure.   For a given ultimate load, alternative arrangements of reinforcement can be assessed for flexural adequacy at the ultimate limit state enabling the designer to optimise/minimise the number of reinforcement configurations, producing a more cost-effective reinforced slab.

Arrangements of reinforcement, for which the corresponding slab ultimate moments of resistance are known, are selected and then checked using analysis software.

It is hoped to develop the software to enable an assessment of shear at supports, and to check the serviceability limit states of deflection and cracking.

The following developments and applications of the automated yield-line analysis are proposed for investigation on the in-situ concrete building.

- application of the existing software to the analysis of all slab panel types to determine their ultimate load factors
- use of the software to examine alternative reinforcement layouts, since preliminary investigations indicate that significant economies of reinforcement may be obtained by the application of collapse load analysis
- inclusion of collapse shear force and reaction determination within the existing software and consideration of the inclusion of displacement determination at working loads

4.3 Distribution of lateral loads

Important assumptions are made in design about how lateral load is distributed in multi-storey buildings. In particular the effect of cladding is normally ignored.

It is normal in concrete construction to assume that all the lateral load is taken by shear walls and shear wall cores. Frame action in concrete buildings is not normally considered although this is not so in steel construction.

Ignoring the cladding is done because it is not easy to model the cladding stiffness and such a process is conservative. For smaller buildings (say less than 10 storeys) the lateral load provisions do not cause problems and probably do not add much to the total cost of the structure. However, estimation of the actual stiffness of buildings is important in relation to the following:

- the response of the building in buffeting winds in relation to comfort of occupants
- lateral and torsional stability/elastic critical loads
- design for earthquake loading

Lateral stiffness of the in-situ building can be inferred from induced vibration tests. Measurements of bare building stiffness and with cladding systems added would be taken.

Initially only the elastic stiffness and stresses would be measured but later tests might involve loading beyond the elastic limit to assess strength and ductility. The full scale test will be backed up by a programme of scale testing and analytical modelling.

4.4 Whole building behaviour

This project is closely linked with the programme on dynamic behaviour and much of the experimental data will be provided by the tests described in connection with that. The project will concentrate on aspects of behaviour specially concerned with an innovative procedure developed for global stability analysis relevant to the global design of buildings. The studies will include: the lateral and torsional stiffness of the structure with/without steel cross bracing which is designed to be asymmetric in plan; the contribution of floors and columns to lateral stiffness; the effect of cladding and non-loadbearing walls; and mode coupling in 3-D behaviour.

As the location of the shear centre is an important piece of information in any structural analysis regarding global behaviour, the data obtained on this "real" structure (which may well be a model for future in-situ construction) would be of great importance.

Behaviour under vertical load on floors will also be investigated to predict the critical load by applying the Southwell plot. Studies will include correlation with previous work at BRE.

5 Structural performance

5.1 Punching shear and redistribution of moment

With the exception of the 9 panel structure tested by the PCA in the USA in the early 1960's and recent full scale tests at the BCA, few if any full scale tests have been carried out on representative flat slab structures. Most codes are based on the results of full scale tests on isolated components (e.g. edge columns and internal columns) or scale models of one or two panels. Since many of the innovative developments which have taken place over the past 30 years are difficult to test using models the proposal to build a full reinforced concrete structure at Cardington presents a unique opportunity for researchers in this area. The unique features of such tests would be:

- full scale testing in a relatively stable environment
- realistic boundary conditions - PCA tests were flawed in this regard
- opportunities to provide comprehensive instrumentation including internally strain gauged reinforcing bars at strategic locations

The results of these tests could give useful data on:

- redistribution of moments
- comparative performance before and after fire damage
- the effect of the introduction of innovative features such as shearheads on the ultimate capacity of interior, edge and corner columns
- reserves of strength from continuity effects

This is a major project which will also include tests on subassemblies of the in-situ building. the test work will help to check and calibrate analytical methods for flat slab design. This information would be of great value for the relevant sections of Eurocode EC2.

5.2 Holes in flat slabs

Two aspects of this problem are to be investigated. The first one is to examine the structural implications of holes of different size and location and assess the performance of slab with holes in practice in comparison to theoretical predictions. The effect on deflections and cracking will also be considered.

The second aspect is the formation of holes in slabs retrospectively by bonding external steel or fibre composite plates. The objective is to provide clear technical guidance on the design and formation of openings in concrete floors using plate bonding technique and other possible systems.

5.3 Study of differential settlement

A project is planned to look at the effect of jacking up and down a series of columns at ground floor level to simulate the effects of subsidence and heave on a frame structure. A series of jacks will be installed at the bases of a number of columns so that one end and one corner of the building can be jacked up and down. To facilitate the jacking, steel inserts are being cast within the columns during the construction phase.

The objective of this work is to produce in a controlled environment deformations which could affect the behaviour of a real structure and to investigate their effect. Besides providing greater understanding of the global behaviour of the structure, the tests will provide valuable data upon which methods of measurement and control of the response of the structure and the occupants of the building can be based.

5.4 Smart technology

Appropriately applied Artificial Intelligence (AI) techniques can be used to handle data collected from monitoring instrumentation and to make decisions as to courses of action to be followed.

An example of this is in relation to the artificially induced subsidence movements induced in the building where an AI system could be trained to correctly predict what is happening to the structure and what action should be taken.

A further application might be where the structure is the subject of repair and the AI system, handling measurements taken from monitoring instrumentation, could provide warning of potential problems.

The particular advantage of using the concrete building for the development of this technology is that, since it will be a test building, it will be possible to introduce gross deformations and try out techniques which would not be possible in a conventional structure. It might be possible to take the technology a stage further allowing the AI system say to control remedial measures, for example reversing the artificially-induced subsidence movements.

6 Cladding research

The type of cladding to be provided on the in-situ frame will be determined by the research requirements and funding available. Cladding is a major component of a building. It determines the architectural appearances of the building as well as affecting its durability, structural performance and level of comfort. With new technology available, cladding can become an interactive component for the collection of energy, the protection from radiation effects in summer, uniform distribution of light, control of temperature and ventilation.

Clearly, there is an opportunity for research into cladding systems and performance on the in-situ concrete building. Some proposals are being actively developed (e.g. explosions and fixings research). The possibilities are broad and could include:

- weather tightness including rain and air penetration and condensation
- innovative systems and products
- thermal insulation
- fit and tolerances
- fixings and anchorage
- accuracy of erection of complete wall
- impact resistance
- bomb blast
- toughening and laminating
- influence on overall building performance

It is hoped that this paper will encourage the submission of research proposals in these and other related areas.

7 Performance in fire

The FRS proposes to carry out an extensive test programme involving fully grown fires within compartments in the building. These tests will move the science of fire engineering further forward and provide valuable input into the conversion of ENV 1992 1.2 to a full EN. The main objectives are:

1. to provide accurate data on the structural response of complete concrete structures to relatively large fully developed fires

2. to study the thermal response of a compartment bounded by concrete construction to fully developed fire
3. to use the information gathered under 1. and 2. above to contribute to the development of a robust approach to the structural fire engineering design of concrete structures

The fire compartments will be established at a number of different levels in the buildings as follows:

• a large compartment covering approximately half floor area at the ground level: this test will allow observation of the behaviour of high strength concrete columns. A number of columns will have polypropylene fibre in the concrete mix in an attempt to avoid spalling.
• smaller fire compartments of one bay size at 2nd, 4th and 6th levels. The main variables between these will be the amount and pattern of imposed load and the fire intensity. The responses produced are expected to vary from simple deformation in the structure to the formation of plastic hinges.

Within this framework of tests, the following projects are being planned in collaboration with the universities:

• structural response of concrete frames to fire
• thermal response of concrete compartments
• robust fire engineering approach to concrete frames
• behaviour of infill concrete/masonry panels in fire
• behaviour of concrete slabs in fire
• measures of alleviate the spalling of concrete in fire
• behaviour of high strength columns in fire
• assessment of post fire damage to concrete
• elevated temperature and moisture migration

8 Response to gas explosions

The test programme planned for the BRE Large Building Test Facility (LBTF) at Cardington provides a unique and ideal opportunity to carry out full-scale tests on a set of structures whose design and history will be well documented. The in-situ concrete building is the first of these structures offering great potential for research into this important area.

Accidental explosions have been shown to cause both considerable hazard to public safety and extensive damage to buildings. The safety of the public within and proximate to such buildings is influenced by the way in which the structure and its individual components perform during, and subsequent to, this extreme form of loading. Once the immediate chaos relating to the explosion itself has passed and public safety has been assured, the building will require refurbishment. If the cost of refurbishment is predicted to exceed or approach that of demolition and rebuilding then the latter option is likely to be selected. The cost of replacing the cladding and services will far exceed the cost of the structure. Thus, improved design of these elements to ameliorate the effects of accidental explosions will result in significant financial benefits while at the same time improving safety. In part, these financial benefits may be due to reduced downtime and consequent loss of revenue to the building occupier and his insurers.

It is clearly desirable to improve the structural performance of cladding and glazing systems under such extreme loads, but without incurring unnecessary and unjustifiable additional cost, either during the construction of new buildings or when improving existing buildings. To do so will require an improved understanding of the way in which an explosive load acts upon, and interacts with, a real building and its neighbours, as well as the way in which the various components interact with each other when loaded in a manner for which they may not have been designed. The objectives of the study are:

- to evaluate the effectiveness of building products designed for improved performance under blast loading, relative to conventional practice.
- to improve understanding of the way in which real buildings (ie., clad, glazed and otherwise equipped) respond to blast loading, how this behaviour can be predicted and how to design for improved survivability. The study will include:

  - blast wave propagation around and inside the structure, in 3 dimensions,
  - response of individual and grouped glazing and cladding panels, including post-failure behaviour such as debris dispersal,
  - transfer of load through fixings to the structural frame,
  - response of the frame.

- to develop improved analytical and design models of the various aspects of building response to blast listed above. These models will be used in the development of improved building products and design philosophies.

This work will be extended to the hybrid and precast buildings.

9 Other research programmes

This paper has concentrated on performance related studies which may be carried out on the in-situ concrete building. There are several other major research programmes which could benefit from this facility. These will of course, need to be developed in detail but brief details are given in the following section.

9.1 Repair and strengthening
Repair and strengthening of concrete structures is a growing field with several new products and techniques being pushed onto the market place. Clearly, these need to perform adequately from the point of view of both strength and long term durability. The in-situ building offers an opportunity for assessing the strength of these repair

systems with reference to framed concrete structures. Some of the topics which would be investigated include:

- retrofitting and strengthening of concrete slabs with large openings using steel or composite fibre plates
- strengthening of column-slab connections (sometimes necessary in car park structures)
- repair and strengthening of damaged columns

9.2 Demolition and recycling

Research proposals in this area may include the following:

1. lightweight aggregate concrete in the steel building
2. innovative techniques for demolition of concrete buildings
3. processing of concrete for recycling
4. uses of recycled concrete
5. use of recycled aggregate in new buildings
6. structural/strength tests
7. research relating to recycled aggregate as material

10 The way forward

It is proposed to establish Common Interest Groups to facilitate the planning and carrying out of the post-construction programme on the in-situ concrete frame. A Common Interest Group (CIG) is designed to bring together people/organisations who could derive mutual benefit from collaborating in research, demonstration or development projects.

For any particular collaboration the aim is to enable the partners involved to invest a relatively small amount of effort and finance and derive a highly geared return. A viable collaboration is achieved by a number of organisations each with their own independent and non-conflicting objectives and funded from different sources having a common interest. The advantage of this form of collaboration is firstly that the cost is spread between a number of organisations and secondly if a single element fails to secure the necessary funding the project (test) as a whole can probably still proceed.

Common Interest Groups can be established most effectively by use of workshops to which those in industry, government and academia who may be interested in undertaking projects within a broad subject area will be invited. Subject areas for initial Workshops relating to the in-situ concrete building at Cardington are envisaged as follows:

- Serviceability and Ultimate Load Performance
- Fire
- Explosions and Impact
- Partitions and Cladding

- Building Services
- Assessment, Repair and Strengthening
- Dismantling and Recycling
- IT, Automation and Robotics

Those with an interest in any of the topics described in this paper or other related studies are invited to join these groups and assist in the planning of the research. By proceeding in this manner, the opportunity offered by this unique and exciting project will be maximised.

11 References

1.   Armer, G.S.T., and Moore, D.B., *Full-scale testing on complete multi-storey structures*, The Structural Engineer, Volume 72, No.2, 1994, pp 30-31.

PART FIVE

Timber Building

TIMBER FRAME 2000

P.J. STEER
Director, P J Steer Consulting, UK

1 Introduction

In 1991 changes to the England and Wales Building Regulations with regard to fire allowed the height of timber framed constructions to be increased from 7.5m to 20.0m, i.e. from 3 storeys to 8 storeys. In the last four years some fifteen timber framed structures have been built either four or five storeys high.

As no specific design rules exist in the UK for these taller buildings, a joint Building Research Establishment (BRE) and TRADA Technology Ltd (TTL) study [1] was commissioned by the Department of the Environment (DoE) to examine how the following aspects were being considered by the building designers:

- strength and stability for static and wind loads
- disproportionate collapse (accidental damage)
- differential movements between the timber frame and the masonry claddings, lifts, stairs, services, etc
- options for structural arrangements of floors and walls with their connections
- fabrication of the frame and construction of the frame and claddings
- services
- building maintenance
- thermal and sound insulation
- quality assurance
- construction regulations with particular reference to fire safety

This study identified areas in the design, fabrication and construction of these higher buildings where no research data existed. To overcome this a research programme was proposed including the construction and testing of a full-sized building.

Consequently, in mid 1995, Timber Frame 2000 (TF2000) was set up, jointly led by BRE and TTL, with the object of ensuring UK plc's position as a world leader in the innovative development of medium rise timber engineered buildings.

2 Timber Frame 2000

TF2000 is a three year programme with the first year (Phase I) devoted to the identification of problem areas requiring testing or verification either by key component tests in the first year, or as part of the 'test bed' building in years two and three (Phase II).

Funding of TF2000 is a joint exercise between the DoE and industry on a 50/50 basis. Part of the work in the first year, which has guaranteed funding from the DoE, is obtaining the necessary support from industry for the development and testing required in Phase II.

The timber frame industry of the UK are major participants in TF2000 representing the different approaches necessary because of different building legislation in England/Wales, Scotland and Northern Ireland. Together with representatives from BRE and TTL they comprise the Management and Technical Committees of TF2000. In due course the membership of TF2000 will be expanded to include material and component suppliers.

TF2000 is administered by a Management Committee whose brief is to:

- define key participants in the project
- define the key technical considerations for further investigation
- establish the funding for the projects and the financial control thereof
- organise publicity, promotion and the development of marketing opportunities
- develop, in Phase I, the construction and test programme for Phase II
- organise and award the various contracts for construction in Phase II
- maintain contact with similar international projects.

The Technical Committee is responsible to the Management Committee for all the technical aspects of the project. These include:

- demonstration of the adequacy of the form of construction proposed with regard to structural strength and stability in normal circumstances and in the situation of disproportionate collapse.
- assessment of dynamic response particularly regarding floor and wind loadings
- sound insulation
- fire safety
- harmonisation and rationalisation of the relevant statutory building regulations in the United Kingdom.
- development of construction methods to minimise the effect of differential movement throughout the building arising from the drying out shrinkage of the timber structure in service and possible simultaneous expansion of other material.

- development of construction techniques for these higher buildings with particular reference to the Construction (Design and Management) Regulations 1994 [2].
- development of cost guidance techniques for timber pre-fabricated construction
- preparation of technical reports and specifications for general application to future buildings.

3 Key objectives

The key objectives of the project are:

1. increasing value of the whole design/construction process to the Client by 30%.
2. reducing construction time by 50%.
3. reducing the cost in use by 20%.
4. achieving regulatory harmonisation.
5. improving quality through off site prefabrication and manufacture.
6. publication of a 'National medium rise timber frame building specification'.

The measurement of the improvements given above are to be related to existing constructions of similar size. These comparisons must, of necessity, be with steel, masonry or concrete structures, as a 7 or 8 storey timber framed building has not yet been built. Independent comparisons of the costs and construction times are essential in this regard.

The increased value to the Client can be achieved by a re-assessment of the design and construction procedures. Reduction in construction time and cost reduction are contributors to this increased value.

The statutory building regulations in England and Wales [3], Scotland [4] and Northern Ireland [5] are three separate documents that in broad terms are in agreement but in essential detail with regard to this project can differ. Certain of the present regulations can only be met by introducing masonry and concrete loadbearing 'heavy and wet' constructions that significantly slow down the 'light and dry' timber frame. Rationalisation of the relevant principles and details in the building regulations would be of benefit to timber frame construction in general.

'Prefabrication' and 'standardisation of construction' have unfortunate connotations with the industrialised building boom in the UK during the 1960s. The objective of TF2000 is to establish 'methods of construction' rather than 'systems' so that once a preferred method of assembling a component or joining together such components is established then the building designer is still free to set the size and proportions of building elements and the building overall. Some modification of present timber frame construction details will be inevitable in these higher buildings.

The results of the TF2000 project will be made public including, in particular, any negative results! Actual construction of 4 and 5 storey buildings is happening now and it is anticipated that higher buildings will be considered if not built at the same time that TF2000 is running. Output from TF2000 through the technical press will therefore need to be a continuous process through to the final report on the project and it is hoped a draft 'code of practice' for the construction of medium rise timber frame buildings.

The spin off from the project will undoubtedly benefit other forms of timber frame construction if only from the greater understanding of certain aspects of construction that have developed as 'folk lore' with the passage of time.

4 The way ahead

TF2000 has been operational since October 1995, and has established Focus Groups to consider, resolve and make proposals on the various topics that have been briefly described in the foregoing, e.g. differential movement, acoustics, regulatory issues, performance testing, construction process, architectural/client view, engineering design, etc. The outputs from these Focus Groups will be drawn together to produce, monitor and report on the programme for Phase II. Resolution of the form of the prototype building, to be constructed at Cardington, is scheduled for completion by March 1996. From this the likely overall cost of Phase II can be determined and subject to funding agreements, the detailed design will proceed with a view to starting construction in October 1996. The results of the Phase I research programme are due by August 1996 and will be incorporated in the 'test' building.

It is anticipated that the prototype building will approach the 20.00m height allowed under the England and Wales Building Regulations. It will be a masonry clad construction built entirely in 'light and dry' timber frame incorporating a lift and services. The building will be fitted with windows and doors throughout and those parts of the building to be tested will be fully finished (including decoration where fire testing occurs).

Before and during construction the dimensional accuracies of components and the assembly will be monitored, and, during the two year programme after construction, relative dimensional movements throughout the building will be measured.

Testing will include structural assessment of the whole building before and after constructing the masonry cladding, floor vibration measurements, sound testing, removal of critical parts of the structure to simulate disproportionate collapse and finally fire testing of compartments.

At the conclusion of this three year programme the knowledge gained will benefit all aspects of the design process, manufacturing and construction. It will define the height limitation of present day practices and technology and identify the changes required to progress higher towards the eight-storey limit.

5 References

1.   *Medium rise timber frame buildings - Disproportionate Collapse and other Design Requirements* - Enjily V., Building Research Establishment, and Mettem C.J., TRADA Technology Ltd, March 1995.
2.   Construction (Design and Management) Regulations 1994, SI 1994/3140.
3.   Building regulations (amendment) regulations 1995, SI 1995/1356
4.   Building Standards (Scotland), amendment regulations 1994, SI 1994/1266
5.   Building Regulations (Northern Ireland) 1994, SR 1994/243

INNOVATION – THE CHALLENGE FOR TIMBER FRAME 2000

C.J. METTEM, G.C. PITTS
TRADA Technology Ltd, UK
V. ENJILY
Timber Structures Division, Building Research Establishment, UK

1 Introduction

In the last four years, some fifteen timber-framed structures have already been built at heights of four storeys plus basements or full five-storeys in timber, yet no comprehensive design rules exist in the UK for such buildings.

A co-operative Building Research Establishment (BRE) and TRADA Technology Ltd (TTL) study, funded jointly with the industry through the DoE Partners in Technology programme, was consequently commissioned, for which the feasibility study was completed in 1995 [2]. Recently, they have embarked upon Stage 1 of an exciting full-scale project, entitled Timber Frame 2000 (TF2000). This is intended to generate authoritative guidance which will ensure uniformity of safety, and explore all the best design features and fabrication practices. Incontrovertible evidence will be obtained throughout the project by designing, constructing and testing a full-sized building.

2 Opportunity

Reduced costs, rapid construction times and high build quality are all reasons for the growing interest in timber frame for medium-rise construction throughout Europe. The high levels of automation that can be engineered into the construction process, and the consequent production efficiency gains, environmental benefits and waste reduction, all contribute to making timber frame an attractive framing method for a wide range of buildings in the 5 to 8 storey range. This type of building is ideally suited to many residential, commercial, hotel, office and leisure applications.

3 Feasibility

Improvements in the England and Wales Building Regulations with regard to fire, mean that buildings of more than three storeys are now fully permitted, rather than requiring special waivers [1].

The alterations, which occurred in 1991, allow timber-framed buildings of up to 20 metres to top floor level to be constructed with fire separating components framed in timber, provided that, in conjunction with their linings, these components have a fire resistance of at least 60 minutes. Timber separating floors and walls with this degree of fire resistance have long been a well-established fact of life for the timber frame industry. Actually, there is test evidence that using similar techniques, 90 minutes resistance can be achieved. This would be required in timber frame buildings taller than eight storeys, and there is now no limit in the Regulations as to the height of construction permitted.

4 Innovation

Nowadays, it is almost impossible to read papers and journal articles on new design and construction projects, without encountering the word "innovation". The TF2000 project will undoubtedly be innovative. The list of 'firsts' that are likely to be claimed will include the following:

- The 'first' full-scale prototype timber frame building above four storeys to have been tested to prove conformance with the "disproportionate collapse" requirements of the Building Regulations
- The 'first' timber structure of such size in Europe to be fully assessed as satisfying the requirements of ENV Eurocode 5
- For those concerned with the Scottish Regulations, a 'rule breaker' of perhaps twice the number of storeys than at present practicable, within the restrictions. A structure having compartment floors and fire resistant shafts constructed from so-called 'combustible' materials.
- A unique test-bed for many serviceability design enhancements – good acoustic and vibrational performance, as well as stiffness under static loads etc
- One of the tallest timber-framed buildings for dwelling purposes to be constructed in the UK, and possibly in the world. Also the 'first' large timber frame test building for the Large Building Test Facility

5 Options

It is necessary to pose the question "to what extent will TF2000 be of a genuinely innovative structural form?"

Three or four broad categories of structural form are recognised by architects as having potential for the general design of timber dwellings (there may be others that we have not thought of, and 'bespoke' projects of a special client/architect relationship are excluded here). Those normally listed [3] are as shown in Figure 1 and as follows:

- Post and beam
- Balloon frame
- Platform frame
- Volumetric construction

6 Structure

There are key structural factors which should influence the choice of structural form:

6.1 Overall stability
This is dependent primarily upon the building's 'footprint' in relation to its total mass, and the technology that can be harnessed to mobilise any necessary anchorages. Post and beam, balloon frame and platform frame can all achieve the required overall stability with appropriate building footprints that permit the desired architectural layouts.

6.2 Elemental resistance to horizontal forces
A number of five-storey real projects have already been built, and design studies have been carried out on both these and higher schemes. These suggest that elemental resistance using existing panellised factory pre-fabricated timber wall framing is achievable with, at the most, improvements/enhancements upon the existing technology, rather than demanding revolutionary innovation.

6.3 Resistance to accidental events
Low-rise experience [4], assessments on completed medium-rise schemes, and feasibility studies on five to eight storeys, all suggest that this is achievable without major re-engineering. Tying/bridging solutions are expected to be offered, giving effectively 'modified platform frame'. Stabilising the brick cladding, if this is the selected material (according to customer preferences) for the full eight storeys, may prove more of a challenge than showing that the timber frame itself can resist blast effects from internal and external accidents.

6.4 Normal vertical load-carrying capacity
Feasibility design studies, and existing constructions of up to five storeys have both shown that this is undoubtedly easily achievable.

7 Construction

Of equal importance are the constructional factors:

Top: Platform frame, site built; Platform frame, small prefabricated panels
Centre: Platform frame, large prefabricated panels, balloon frame
Bottom: Volumetric, post and beam

Figure 1.   Timber Frame Techniques

7.1 Vertical differential movements

Need to be either 'designed for' or 'designed out'. These may take place between the timber frame and the 'hard' parts of the construction, i.e. cladding, lifts and services; and between those parts of the frame subjected to various moisture content cycling regimes through the seasons of the year. The particular strengths and weaknesses of the various general forms of structure (platform versus balloon frame for example) are especially worth considering in this respect. The choice of cladding will also be very influential in finally deciding how differential movements will be dealt with.

7.2 Acoustical performance

Greater consideration of optional (non-regulatory) customer needs is drawing attention to the desirability of achieving excellent acoustical performance, both within and between the timber frame compartments, and in terms of the whole building. High performance sound insulation is offered by a number of existing timber floor and wall solutions.

The project will be examining how the whole structure and envelope performs. Vertical flanking sound transmission across a number of flats or apartments is of special interest in this respect.

7.3 Fire safety

Fire will be another important consideration. It is in this area that the overall aim of establishing greater harmony between the regulations of England and Wales, and those of Scotland, takes on special significance. This is because the Scottish Consortium of Timber Frame Industries (SCOTFI) regard the present Technical Standards of the Scottish Building Regulations, when compared with those for England and Wales, as unnecessarily restrictive, and a discouragement to developers and builders. This is especially ironic when it is recalled that in low-rise timber frame construction, Scotland has always led the way in the UK in terms of market share; appreciating its energy efficiency, high comfort levels and competitiveness.

The DoE has recently launched a document entitled "Timber 2005 – A Research and Innovation Strategy for Timber in Construction". This clearly highlights the need for regulatory harmonisation, so that timber products can compete on equal terms with other construction materials. A number of 'all timber' fire safety measures will be incorporated in the prototype, in order to pursue such aims.

8 Credentials

It should not be left unsaid that timber frame construction, irrespective of its particular form (platform frame, post and beam etc) can call upon an excellent track record in terms of well established credentials:

- Excellent thermal performance
- A dry construction process
- An economical and amenable technique for factory pre-fabrication
- Lightness, giving low erection costs and needing only inexpensive foundations
- A good environmental image, backed up by demonstrable whole-life audits

9 Evolution rather than revolution

TF2000 is likely in practice to proceed as an evolutionary, rather than a revolutionary project – *A Concorde rather than a Mosquito!*

This conclusion can be reached by further discussing the above options for structural form:

9.1 Post and beam

Long lengths of laminated veneer lumber (LVL) [5], glulam [6] parallel strand lumber (PSL) etc (up to 24 m) are available. As well as being obtainable in large sizes, these materials have the additional merits of being of high strength and stiffness, 'ultra-dry' and of great dimensional precision. Obviously, their raw material cost per cubic metre is greater than that of dry strength graded solid sawn softwood (current industry "best practice").

What are the total cost implications of designing exclusively with Structural Timber Composites? Are there benefits to be gained from constructing with a mixture of dry strength graded softwood and composites? – We intend finding out!

Could we see full-height, continuous timber columns, leading to a structure of the form of the eight-storey steel-framed building? This would be returning to well established English timber framing principles of Mediaeval times!

9.2 Balloon frame

Balloon frame construction might be adaptable to multi-storey situations. Information is at present being acquired through research on the performance of wall diaphragms of greater than single-storey height. The driver for this is non-dwelling constructional applications, but medium-rise dwelling applications may spin off.

Perhaps Structural Timber Composites could be used in conjunction with balloon frame to achieve innovative forms of timber frame construction?

9.3 Platform frame

This is likely to be the selected option – i.e. evolutionary rather than revolutionary, as indicated above! Figure 2 provides an indication of the nature of the elevations, sections and floor plans being considered at the time of writing. Figure 3 shows, for an indefinite number of storeys, the general manner in which platform frame construction is built.

9.4 Volumetric

This has been used with considerable success for timber-framed construction of up to four storeys. Its main applications are those where there are very standardised client requirements, such as hotel extensions, motels, nursing homes etc. Volumetric construction tends to be too restrictive in floor layout terms for flats and apartments.

10 Focus

TF2000 has established Focus Groups to resolve and propose optimal approaches to the following:

- Strength and Stability of 5 to 8 Storey Timber Engineered Buildings
- Client and Architectural Requirements
- Regulatory Issues and pan- UK Harmonisation
- Achieving Very Robust Response to Accidental Damage
- Ensuring Fire Safety
- Serviceability Aspects likely to Improve Client Satisfaction
- Measurements of Reductions in Construction Time and Cost in Use

Performance testing requirements, and their influence upon the design of the prototype will be settled rapidly by the Focus Groups and a Plenary process. Hence the form of the prototype building is expected to be known very soon. The likely overall costs of its construction will then be calculated, and, subject to funding agreements, the detailed design work will begin. A rapid start on the construction is then scheduled. The results of an initial component testing and desk-top study on close details of the design process, including issues that relate to the use of the new structural timber Eurocode 5, are to be fed in at that stage. This work is already well underway.

It is anticipated that the building will incorporate a lift and services. It will also be fitted throughout with windows and doors. Those parts of the building to be tested for aspects such as acoustic performance and fire will be fully finished internally. The dimensional accuracy of all components and the complete assembly will be monitored throughout the two-year test programme, and relative dimensional changes will also be continuously measured. Testing will include structural assessments of the stiffness of the whole building before and after constructing the cladding; floor vibration measurements; acoustic tests; removal of critical parts of the structure to simulate accidental damage and to prove that there is no risk whatever of disproportionate collapse, and finally fire testing on major compartments.

11 Market pull

Does it actually matter if the scheme seeks to establish incremental improvements rather than being radically innovative?

The project TF2000 is funded through the DoE 'Partners in Technology' programme. It is genuinely collaborative in nature, and is in fact the largest joint venture Research and Technology Development (RTD) project ever embarked upon by the UK timber frame industry. Two leading English companies, and two equally leading Scottish firms, are central to not only the technical efforts, but also a major share of the fund raising.

Nowadays all innovative products must be relevant to the market's demands. Any necessary and associated RTD must be linked to the industry's needs. The products of 'technology push' are not always welcomed. Researchers who exhibit even a hint of wishing to work on problems that merely interest them, are regarded with the greatest disapproval!

If the needs of tomorrow's markets can be satisfied by 'evolutionary innovation', then so well and good!

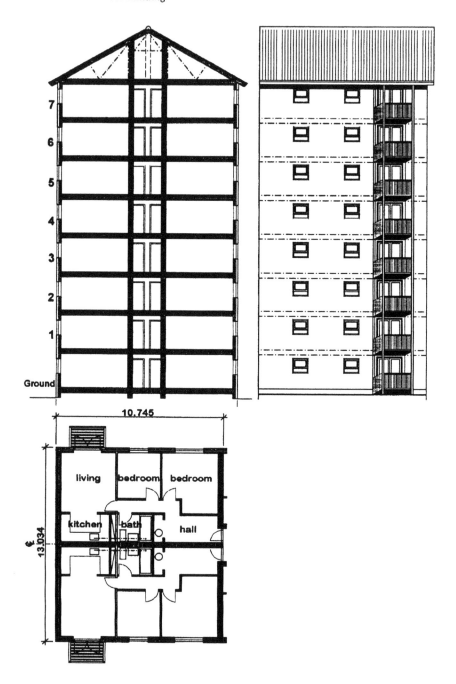

Figure 2.   Medium Rise Timber Frame Building – Section, Elevation, Typical Flat

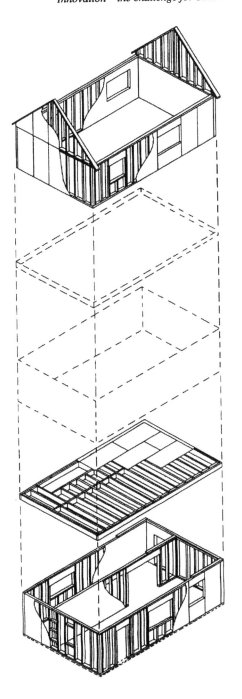

Figure 3.   Platform Frame Construction

Fully developed solutions and construction methods will give timber frame access to new and extended markets, including:

- New homes (worth £6.5 billion annually)
- Commercial offices (worth £2.1 billion annually)
- Student accommodation (worth £350 million annually)
- Hotels, Hostels, Nursing homes, Schools, Barracks etc.

There is also a growing overseas market for factory-engineered building solutions, and the export potential is enormous. International contacts established during the feasibility study confirmed that British companies are in a leading position in having constructed a number of real projects of four and five storeys in timber frame. These are regarded as very adventurous by others in the field, including Scandinavia, North America and several Southern Hemisphere groups.

It is expected that Project TF2000 will revitalise interest in this highly efficient method of building. It will provide a platform from which developers and manufacturers will launch many new products and concepts.

The Scottish timber frame industry is extremely keen to demonstrate that harmonisation of its Regulations with those of England and Wales will remove unnecessary restrictions, and eliminate a reduction of choice for developers and builders.

The timber frame industry in the whole of the UK is convinced that its clients will appreciate the excellent energy efficiency, high comfort levels and competitive build costs of taller timber frame buildings. The importance of this project, and the positive publicity that it will attract, will ensure the highest possible levels of interest in the businesses of all the participants.

Project TF2000 will undoubtedly underpin British wealth creation, and will further enhance our competitiveness in this field.

12 References

1.   Steer, P.J. *Timber Frame 2000*. Second Cardington Conference, 1996.
2.   Enjily, V. & Mettem, C.J. *Medium Rise Timber Frame Buildings – Disproportionate Collapse and other Design Requirements – A Nine-Month Feasibility Study*. Joint BRE/TTL Final Project Report for DoE and Industrial Partners, 1995.
3.   TRADA Technology Ltd. *Timber Frame Construction*. ISBN 0 901 348 94 5. 2nd Edition 1994. High Wycombe.
4.   Mettem, C.J. & Marcroft, J.P. *Simulated Accidental Events on a Trussed Rafter Roofed Building*. CIB-W18/26-14-6. Athens, Georgia, USA, 1993.
5.   TRADA Technology Ltd. *Structural Timber Composites – A Generic Guidance Document. In Course of Preparation*, 1996.
6.   Glued Laminated Timber Association. *Specifiers' Guide to Glued Laminated Structural Timber*. Chiltern House, Hughenden Valley, High Wycombe, 1991.

Discussion Paper

DYNAMIC RESPONSE TESTING AND FINITE ELEMENT MODELLING OF THE LBTF STEEL-FRAMED BUILDING FLOORS

A. ZAMAN and L.F. BOSWELL
Department of Civil Engineering, City University, London, UK

1 Introduction

Floor vibrations are becoming increasingly important for both serviceability limit states and safety requirements. This is mainly due to the increasing trend towards lighter and longer span floors in all forms of construction. Composite floor vibrations caused by internal sources have previously been studied and a design guide [1] produced. However, this guide is not based on experimental results and provides only a conservative design method for assessing the vibrational behaviour of floors in steel framed buildings only. Thus, there is a need to study experimentally the behaviour of these floors.

An extensive experimental programme to examine the dynamic behaviour of a range of different floor slabs have already been completed at City University. However, this paper only describes the information derived from and the results of full scale dynamic tests carried out on the LBTF building floors at Cardington. Two types of tests were carried out: hammer and heel drop. The experimental results were validated by a finite element (FE) model of the floor. It is shown that a single-panel FE model may be used to predict the fundamental natural frequency of these types of floors. The model can be used to ensure safe frequencies and deflections for serviceability limit states by varying the various parameters.

2 LBTF building floors

Figure 1 shows the plan layout with the test grid and a typical cross-section of the LBTF building floors. The floors are made of 130 mm thick light weight concrete (1900 kg/m^3 [2]) of grade 30 with CF70 profile [2] (Figure 1b) supported on steel beams of grade 50. The modulus of elasticity of the concrete was assumed to be 22

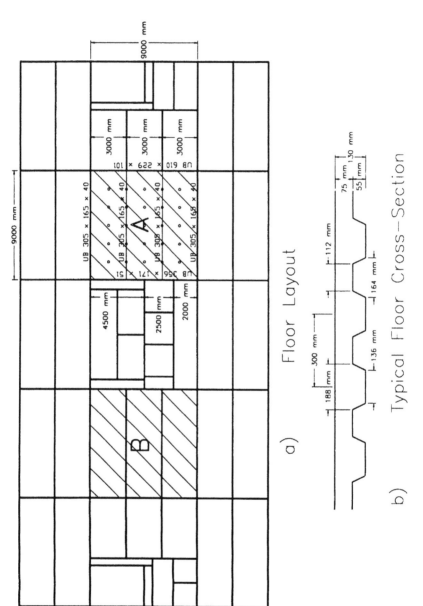

a)   Floor Layout

b)   Typical Floor Cross-Section

Figure 1.   Steel-framed building at the Cardington LBTF

Gpa [1] whereas the density and elasticity modulus for steel were fixed in the I-DEAS finite element software to be 7820 kg/m³ and 206.8 GPa, respectively.

3 Dynamic testing

Previous dynamic testing have focused on the lateral vibrations of the whole LBTF building using laser measurements. The dynamic behaviour of the building floors, however, have not been considered so far. The main objective of the tests on the Cardington floors, therefore, was to determine the dynamic characteristics of the fundamental mode of vibration of the floors and the response of floors to a sudden impact.

Three floor panels (two at level-5 and one at level-3) were tested. The two level-5 panels had earlier been subjected to different loading tests but were unloaded at the time of the tests. Both the panels were tested since the visible amount of cracking was different for the two panels. The level-3 panel was loaded with serviceability load of 1000 kg sand bags placed uniformly on the floor. The tests were carried out on April 3 1995.

3.1 Hammer Testing

Hammer testing [3] was used to measure the fundamental natural frequency and damping of the floor. It is the simplest, quickest, easiest, least expensive and most portable method available with a compact kit. The test is non-destructive and takes on average about 2–4 hours. It is assumed, however, that the test floor is linear and excited only in its linear range.

The test involves exciting the floor by a soft-tipped 5.4 kg hammer instrumented with a force transducer and measuring the response with an accelerometer. A spectrum analyser is used to extract this information from the hammer signal and compare it to the signal generated by the accelerometer located at various points on the floor. The analyser carries out the Fourier transforms and displays the inertance transfer function, phase and coherence which are used in later analyses to extract the dynamic parameters of the floor.

The test commences by marking a grid of points on a typical panel of the floor. The most flexible panel is normally selected to measure the lowest natural frequencies. The accelerometer is placed on a grid point that gives good response at the natural frequencies of interest (i.e. not near a nodal line). This is done by quickly testing a few points on the grid. Normally, the midspan point gives the best response due to relatively larger amplitudes and the accelerometer is placed there for measurements. The floor is then excited by the hammer at all the grid points. Alternatively, the accelerometer may be moved on the test grid while exciting the floor at the midspan location. A minimum of five impacts are used at each grid point to obtain an average value of the response at these locations. The spectrum analyser performs the Fourier analyses and provide plots of transfer function, phase and coherence function. A frequency band-width of 25 Hz may be used on the analyser to estimate the first few frequencies of the floor. Data for each grid point, measured in the above manner, is stored on the analyser for later transfer to a PC for further analyses to extract the dynamic properties. A multi-degree-of-freedom curve-fit program is used to extract frequencies and damping from the peaks of the transfer functions for each grid location.

3.2 Heel Drop Tests

Heel drop test [1] was used to measure the vibration response to a sudden impact. This test is performed by a person of average weight, standing at the centre of the floor panel, rising onto the balls of his feet and then dropping down on his heels. The resulting acceleration time history is measured by the accelerometer placed near the feet of the test person. The time history gives the initial peak acceleration due to the heel drop. Since the force input cannot be measured in this test, the results may vary from subject to subject. However, a response spectrum of the time history provides good estimates of the frequency and damping of the floor.

4 Experimental results

Figure 2 to Figure 7 shows the typical hammer and heel drop test results for the three panels of the LBTF building floors tested, respectively. The coherence function is only an indicator of the quality of measurements.

Table 1 and Table 2 gives the hammer and heel drop test results for the three panels tested. The heel drop peak accelerations have been corrected for the initial amplitudes.

Table 1. Hammer test results of LBTF building floors

Floor	Fundamental Natural Frequency (Hz)	Damping (% critical)
Level-5 (A)	6.39	3.61
Level-5 (B)	6.26	3.60
Level-3 (A)	4.66	7.88

Table 2. Heel drop test results of LBTF building floors

Floor	Fundamental Natural Frequency (Hz)	Damping (% critical)	Heel Drop Acceleration (% g)
Level-5 (A)	6.38	3.41	9.19
Level-5 (B)	6.13	4.37	7.42
Level-3 (A)	4.50	1.91	2.32

4.1 Fundamental natural frequency

The measured fundamental natural frequency of the bare composite floor panel-A at level-5 was found to be slightly higher than that of panel-B at the same level. Even though both the panels had visible cracks, the main reason for the difference in this frequency was due to the additional cracking of panel-B floor. This reduced the effective moment of inertia and thus the floor frequency. The level-3 panel-A had additional load of 1000 kg sand bags placed uniformly on the floor. This additional mass reduced the frequency by about 27.07%.

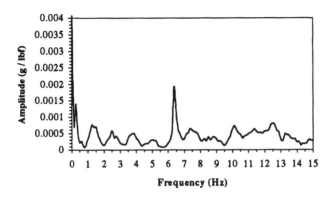

a)   Transfer Function

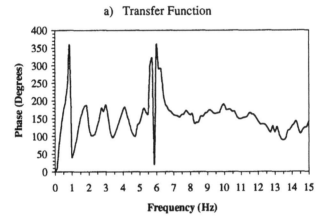

b)   Phase

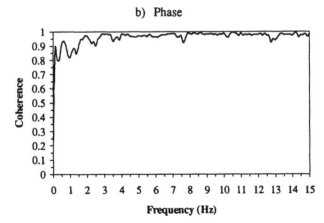

c)   Coherence Function

Figure 2:   Typical Hammer Test Results for Level-5 (A)

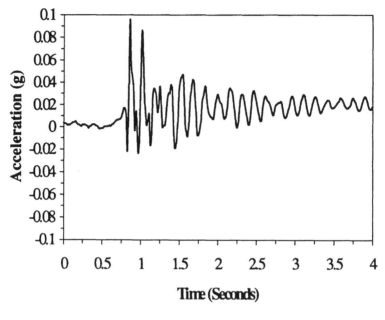

a) Time History

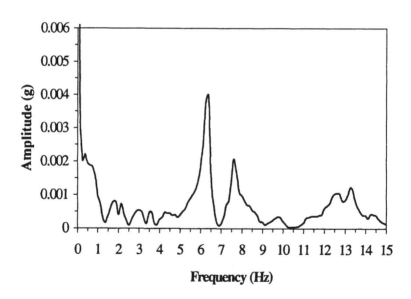

b) Auto-Spectrum

Figure 3: Typical Heel Drop Test Results for Level-5 (A)

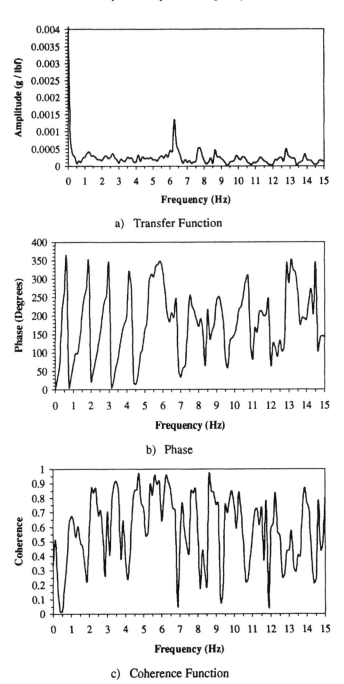

a)   Transfer Function

b)   Phase

c)   Coherence Function

Figure 4:   Typical Hammer Test Results for Level-5 (B)

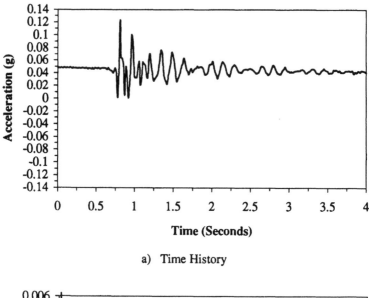

a) Time History

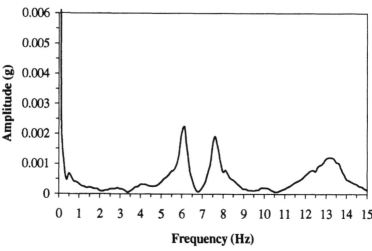

b) Auto-Spectrum

Figure 5:   Typical Heel Drop Test Results for Level-5 (B)

The fundamental natural frequencies obtained from the spectrum of the heel drop time histories shows good correlation with the hammer test results. The heel drop spectrum also clearly identifies the various peaks which is difficult to identify in the case of some hammer test results due to pollution caused by noise at the time of the test.

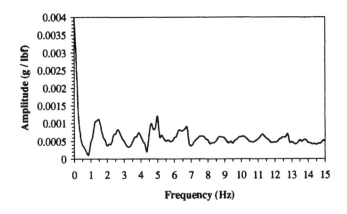

a) Transfer Function

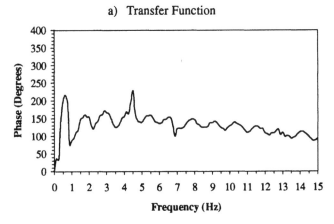

b) Phase

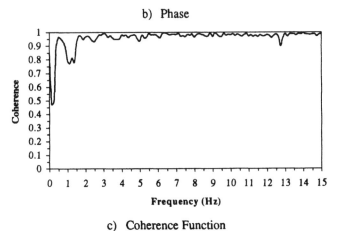

c) Coherence Function

Figure 6: Typical Hammer Test Results for Level-3 (A)

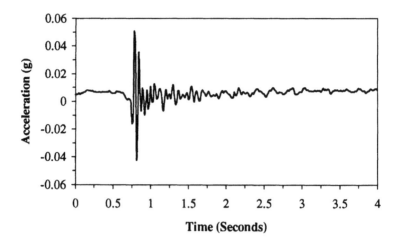

a) Time History

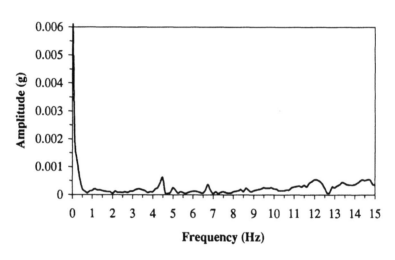

b) Auto-Spectrum

Figure 7:   Typical Heel drop Test Results for Level-3 (A)

4.2 Damping

There was negligible difference in the damping values of level-5 panels. However, the additional mass at level-3 panel apparently increased the effective damping.

Damping is measured by the half-power band width method which only considers the peaks in the transfer functions or auto spectrums. Since peaks can only be identified clearly by studying the corresponding phase information, it may be possible to estimate wrong damping values for two adjacent peaks, or where a transfer function is polluted with noise etc. Therefore, the damping estimates for the hammer test results for level-3 panel may be assumed inaccurate. This also emphasises the fact that whereas frequency estimation, using either test procedure, may be accurate, damping values may be unreliable in real life structures. Note that the damping values for the hammer test results are based on an average of a large number of total impacts at various locations on the floor. The heel drop damping estimates, however, are due to a single impact at or near the midspan location only.

5 Finite element modelling

Research [3, 4] has shown that the fundamental natural frequency of a post-tensioned concrete floor can be estimated by a single-panel linear-elastic FE model of the floor. The FE modelling of the LBTF building floors was, therefore, based on this concept. Two models were studied for the composite profiled floor slab. In both, the slab was modelled as shell elements and the steel beams as beam elements with their centroids connected by rigid elements. Actual steel sections from BS5950 (1985) were used for the beam elements. The rigid elements define the correct restraint conditions for the displacement of the welded joints at these locations due to their relatively high local stiffness. Column supports were assumed as pin joints. Figure 8 and Figure 9 show the two models and their fundamental mode shapes.

Model-A The slab was modelled as shells of uniform thickness, calculated as:
a) Based on actual profile dimensions for the dimensions in Figure 1:

$$t = 130 - \frac{112 + 164}{2} x \frac{55}{300} \cong 105 \text{ mm} \tag{1}$$

b) Based on Composite Profile Slab Weight:
   The total weight of the 130 mm deep composite profiled slab CF70 [2] for 75 mm slab thickness of light-weight concrete is 2.07 kN/m². Therefore,

$$t = \frac{2.07 x 1000}{9.80665} x \frac{1000}{1900} \cong 110 \text{ mm} \tag{2}$$

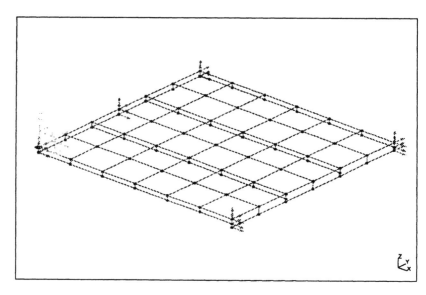

b) Model-A

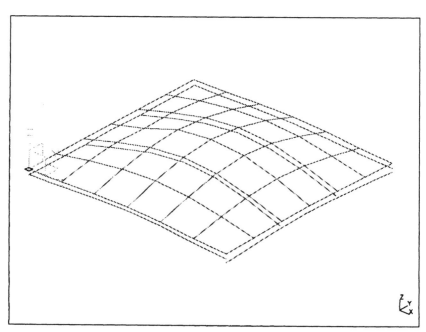

b) Fundamental Mode Shape

Figure 8: Finite Element Model-A of the LBTF Floors

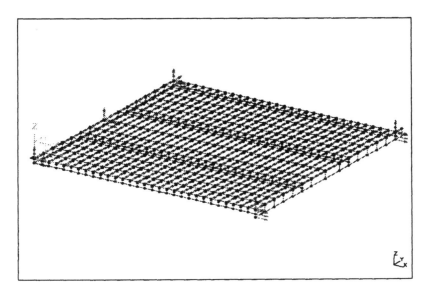

a) Model-B

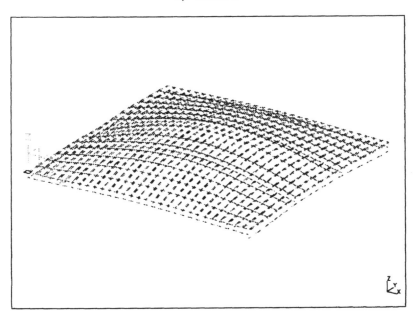

b) Fundamental Mode Shape

Figure 9:   Finite Element Model-B of the LBTF Floors

For a conservative estimate, based on the lower slab thickness of 105 mm, fundamental natural frequency was found to be 7.17 Hz. This is about 12.21% higher than the experimental frequency. The main reason for this high frequency is the uniform thickness for the profiled deck slab. Also, the floor frequency may have been affected due to its reduced moment of inertia because of cracking as a result of various loading tests on the floor.

Model-B   The slab was modelled as shells of 75 mm thickness and the profile as shells in the vertical plane of 55 mm deep and 162 mm thick. This thickness was calculated as average of the actual profile dimensions (Figure 1). The fundamental natural frequency of this model was 6.37 Hz, i.e. 0.31% less than measured. This model represents a more realistic approach as compared to Model-A. However, the model requires a large number of elements and take a longer computer CPU time for analyses. It may be noted, however, that although the average profile dimensions are used, the metal decking is not considered in modelling. This model will, therefore, always result in lower frequencies than experimental and model-A frequencies. The reason for close results in the case of LBTF floors has already been discussed as cracking due to other loading tests.

5 Discussion

The difference in the estimation of frequencies using the two different models is within 15% ±. Therefore, any of the two single-panel models may be used at the design stage to estimate the fundamental natural frequency of these types of floors. The use of such analyses also allows the designer to vary the span and/or slab thickness and steel section for acceptable vibrations.

The fundamental natural frequency of the LBTF floors was above 6 Hz (below which, the NBCC [5] would require a full dynamic analysis) even though the floors had been cracked due to loading tests. Further, the frequency is not in the resonance range of excitation frequency due to various pedestrian activities. Whereas this may discourage any dynamic response analyses or testing, however, the high floor accelerations due to a single impact at the point of maximum deflection, by a human of normal weight, are significant.

6 Conclusion

The results of the heel drop test correlates closely with the hammer test results in estimating the floor fundamental natural frequency. Hammer testing can be better employed where a more accurate estimate of damping and mode shape is desired whereas a heel drop test also provides the acceleration levels for a sudden impact. The tests, however, need to be carried out in a very quiet environment in order to obtain the best results. The information from the performance of the prototype floor to these tests may be used to refine and improve the design of composite floors.

Although a limited number of tests have been conducted, the presence of an imposed dead load at Level-3 floors reduced natural frequencies and increased floor damping, which was expected.

Since frequency alone is the most important factor in controlling structural vibrations, it is, therefore, important to estimate frequency as accurately as possible. The FE modelling described here may be used for this purpose. The single-panel model is found to estimate accurately the fundamental natural frequency of composite floors [3, 4] for post-tensioned concrete floors. Static analyses of the model also allows the design of appropriate steel sections for these floor types and to ensure high frequencies even after the application of heavy loads.

The fundamental natural frequencies of composite floors are normally not in the range of excitation frequencies due to pedestrian movements. Therefore, resonance conditions are not a problem. However, the high floor accelerations due to a sudden impact may cause perceptible vibrations. Although it is very difficult to modify an existing floor to reduce its vibration susceptibility, however, worrying accelerations can be dampened out quickly by providing reasonable partitions etc.

7 References

1.   Wyatt, T.A., *Design Guide On The Vibration Of Floors*, The Steel Construction Institute, UK, SCI Publication 076, 1989, 43 pages.
2.   Precision Metal Forming Limited, *ComFlor 70*, Brochure on Composite Floor Decking Systems, 1994, 12 pages.
3.   Zaman, A., *Frequency Estimation of Pre-Stressed and Composite Floors*, PhD Thesis, Department of Civil Engineering, City University, London, 1996.
4.   Zaman, A. and Boswell, L.F., *Limiting Deflections And Span/Depth Ratios For The Vibration Of Post-Tensioned Concrete Flat Slab Floors*, submitted to the ICE Proceedings Journal.
5.   CSA Standard, *National Building Code Of Canada (NBCC)*, Sentence 4.1.10.4.(1), Canadian Standards Association, National Research Council of Canada, 1985.

8 Acknowledgements

The authors would like to thank Mr David Cobb of BRE for his permission to carry out the testing.

KEYWORD INDEX

www.ingramcontent.com/pod-product-compliance
Ingram Content Group UK Ltd.
Pitfield, Milton Keynes, MK11 3LW, UK
UKHW020433010325
455677UK00029B/1136

9 780367 863708